Fruits of Eden

PROOF

UNIVERSITY PRESS OF FLORIDA

Florida A&M University, Tallahassee
Florida Atlantic University, Boca Raton
Florida Gulf Coast University, Ft. Myers
Florida International University, Miami
Florida State University, Tallahassee
New College of Florida, Sarasota
University of Central Florida, Orlando
University of Florida, Gainesville
University of North Florida, Jacksonville
University of South Florida, Tampa
University of West Florida, Pensacola

proof

FRUITS OF EDEN

David Fairchild and America's Plant Hunters

AMANDA HARRIS

proof

University Press of Florida

Gainesville · Tallahassee · Tampa · Boca Raton

Pensacola · Orlando · Miami · Jacksonville · Ft. Myers · Sarasota

Library of Congress Control Number: 2014953611
ISBN 978-0-8130-6061-3

The University Press of Florida is the scholarly publishing agency for the State University
System of Florida, comprising Florida A&M University, Florida Atlantic University, Florida
Gulf Coast University, Florida International University, Florida State University, New
College of Florida, University of Central Florida, University of Florida, University of North
Florida, University of South Florida, and University of West Florida.

University Press of Florida
15 Northwest 15th Street
Gainesville, FL 32611-2079
http://www.upf.com

For Drew

proof

Contents

Introduction

At the end of the nineteenth century, most food in America was bland and brown.

The typical family ate pretty much the same dishes every day. Their standard fare included beefsteaks smothered in onions (a condiment that passed for a vegetable), ham with rank-smelling cabbage, or maybe mushy macaroni coated in cheese. Since refrigeration didn't exist, ingredients were limited to crops raised in the backyard or on a nearby farm. Corn and wheat, cows and pigs dominated American agriculture and American kitchens.

Farmers grew the limited range of crops that had haphazardly found their way to the New World. Only a few dozen foods—mainly squashes, grapes, berries, corn, and potatoes—are native plants; men and nature had carried the rest from other continents. Most Americans didn't know what a fresh orange looked like or how to peel a banana.

As the century ended, however, Americans opened their eyes to the world beyond their shores. Many things changed during this pivotal period when globalization began. And thanks to an earnest young botanist from the Midwest named David Grandison Fairchild, national eating habits was one of them.

Fairchild transformed American meals by introducing foods from other countries. His campaign began as a New Year's Resolution for 1897 and continued for more than thirty years, despite difficult periods of xenophobia at home and international warfare abroad. After he persuaded the United States Department of Agriculture to sponsor his project, he sent other smart, curious botanists to Asia, Africa, South America, and Europe to find new foods and plants. They explored remote jungles, desert oases, and mountain valleys and shipped their discoveries to government gardeners for testing across America. Collectively, the plant explorers introduced more than 58,000 items.

Supported by his wife Marian's relatives—her father, Alexander Graham Bell, and her brother-in-law, Gilbert Grosvenor, editor of the *National Geographic Magazine*—Fairchild was a member of an influential group whose passion for invention and discovery helped propel America out of the isolation of the nineteenth century into its cosmopolitan eminence of the twentieth century.

While other types of explorers worked with well-financed teams to brave the jungles of South America and the frigid peaks of the North and South Poles, Fairchild and his colleagues worked alone or in small groups. They traveled by ship, train, horse, mule, and on foot. Ironically, only three weeks after Fairchild's star explorer, Frank Meyer, returned from his first walking expedition through China, Fairchild went to Hammondsport, New York, to help Glenn Curtiss test his aircraft, the *June Bug*. It was before the Wright brothers had flown their plane in public, so Curtiss's demonstration was the first time any Americans had witnessed human flight. The breakthrough in aviation changed exploration, for plants and everything else, forever.

Despite the risks and prejudices they faced, Fairchild and his colleagues found foods that continue to enrich the lives of Americans. Many of their discoveries have been used as breeding material to improve existing plants, and others have become staples of the American table like mangoes, avocadoes, soybeans, figs, dates, and Meyer lemons.

And almost all are tasty and colorful.

Escape from Kansas

The creatures swooped down from the high peaks of the Rockies and traveled east at a rate of 20 to 30 miles a day. From a distance they resembled a giant snowstorm darkening the sky. Up close they produced an eerie whirring noise like hailstones hitting the ground.

They were grasshoppers or, more precisely, Rocky Mountain locusts, fierce bugs that devastated farms in Kansas, Missouri, Minnesota, and Nebraska throughout the 1870s. Millions or perhaps billions of them, each less than an inch and a half long, crammed together in a pack that, one expert determined, sometimes covered an area 1,800 miles long and 110 miles wide. They flew so densely packed that their wings tangled together. "They seem to cover the face of the earth," said Willis Parkison Popenoe, a Topeka farmer who had watched them approach.[1]

The insects were starving and frantic to find food. Their jaws worked constantly to devour everything in their path: wheat, barley, and green vegetables. They stripped fields of corn in a few hours and then ate the wooden fences built to protect crops from predators. After they finished the leaves on the trees, they consumed the bark and then destroyed leather stirrups or gloves lying on the ground. They ate clothes hanging outside to dry.

A child wearing a white dress with a green stripe down the back froze in terror as grasshoppers devoured the dyed fabric she was wearing. Another little girl on the prairie tried frantically to brush away the grasshoppers that covered her body. "Their claws clung to her skin and her dress," one account read. "They looked at her with bulging eyes, turning their heads this way and that."[2]farmers. With no understanding of why the invasion had occurred, many concluded that this had to be the wrath of God.

*   *   *

Superstition was common in the small city of Manhattan, Kansas, when George Thompson Fairchild arrived in 1879. He came by train from Michigan with his wife and five children to start a new job as president of Kansas State Agricultural College. The school's goal was to educate students in many ways: to teach them practical skills and instill an understanding that science governed natural phenomena like locust swarms. This knowledge was essential at the time when farming was the social and economic bedrock of American life.

George Fairchild, forty-one years old, was a Puritan who had taught English literature for many years at Michigan State Agricultural College in Lansing. America had only a few public colleges at the time, but Fairchild was already a prominent educator because of his conviction that the whole nation would benefit if farmers' children received solid, well-rounded educations instead of narrow vocational training. "College is not so much to make men farmers as to make farmers men," he said.[3]

Kansas officials were pleased to have recruited Fairchild at a key time in the college's development: the school, which was an important component of the federal government's system of land grant institutions, had become a hotbed of exciting scientific ideas since its opening fourteen years earlier. When Fairchild took charge, the college had 207 students and twelve professors who taught everything they needed to know to succeed in life at the end of the century, from Latin declensions to chemistry.

Fairchild was a high-minded teacher born into a family of educators; his grandfather had been a founder of Oberlin College in Ohio; one of his brothers was Oberlin's current president, and another was a founder of Berea College in Kentucky. These were all progressive schools that accepted women and, in the case of Berea, blacks. Fairchild's wife, Charlotte Halsted, was a devout Quaker who had been one of the first women to graduate from Oberlin. Fairchild was a careful, mild-mannered scholar; in Kansas he took time to prepare individual assignments for each student at the start of every term.

George and Charlotte Fairchild "belonged to a class to whom the intellectual future of this country meant more than anything else in the world," their youngest son, David, observed many years later.[4]

When the Fairchild family arrived in Kansas, two thousand people lived in Manhattan, a fertile swath of riverfront land in the eastern section of the state. After the U.S. government forced native tribes to leave Kansas, abolitionists had arrived from the East in an organized drive to outnumber proponents of slavery. They wanted to guarantee that a majority of Kansas' settlers would vote to become a free state when it was eventually admitted to the union. Antislavery people from New England and Ohio first settled the area in 1855; they originally called it Boston because they had been members of the New England Emigrant Aid Society. Within weeks they changed the name to Manhattan to honor investors in New York City who had helped finance their trip west.

David Grandison Fairchild was the fourth of five children. He was born April 7, 1869, in East Lansing on the campus of Michigan State. He was ten when his family arrived in Kansas, old enough to be shocked by the natural differences between Michigan and Kansas.

His uncle Byron Halsted, a botanist and early authority on plant diseases, had taught him to appreciate nature when he was growing up amid the primeval forests of magnificent oaks, sugar maples, and majestic black walnut trees that surrounded the Michigan campus. Kansas, in contrast, was the heart of the Great Plains, an area of vast prairies almost devoid of trees. Fairchild called his new home "the far west" because the state was on the opposite side of the Mississippi River.[5]

With a Quaker mother and a Puritan father, David Fairchild grew up in a household with many rules: no dancing, no music, and few examples of art, nothing that might overstimulate the senses of growing children. Fairchild "had been brought up to plain living and high thinking," recalled Marjory Stoneham Douglas, a writer who became Fairchild's close friend in Florida.[6] He was an obedient boy who milked the family cow and at the age of ten took the Blue Ribbon Pledge to abstain from liquor for the rest of his life. Adventure stories were among the few exciting books allowed in his home.

"Among the children of my generation, the *Arabian Nights' Entertainments* held a place second only to *Robinson Crusoe*," Fairchild remembered. "The stories were read to us before the open fire on winter evenings, and

Baghdad seemed a dream place somewhere in the dim never-never land near Crusoe's tropical island."[7]

*   *   *

When he was about fifteen, Fairchild enrolled as a student at Kansas College to study horticulture and botany. He made two important friends there: Walter Tennyson Swingle, another botany student, and Charles Lester Marlatt, a young entomology instructor. The three would remain close associates for the rest of their lives, in good ways and bad.

Fairchild's ideas about the natural world widened in May 1887 when he caught a glimpse of life in a distant, exotic land. His friend Marlatt invited Alfred Russel Wallace, a celebrated English naturalist who had discovered the principles of natural selection simultaneously with Charles Darwin, to speak at the college during his tour of America.

Wallace had traveled for eight years over 14,000 miles throughout the South Seas and returned to England with revolutionary insights into evolution. His best-known book, *The Malay Archipelago*, was published in 1869. The eminent visitor was invited to President Fairchild's home for tea one evening during his brief visit to the school.

Wallace's description of his adventures in Java and other remote islands captivated David Fairchild. Wallace probably repeated exploits he had already recounted in his letters and books: how he had carefully nursed an orphaned baby orangutan in Borneo, tracked down the near-mythical but real bird of paradise, and admired the giant ferns on Dobu Island. Those plants were "trees bearing their elegant heads of fronds more than 30 feet in the air," Wallace wrote in *The Malay Archipelago*. "There is nothing in tropical vegetation so perfectly beautiful."[8]

Wallace may have declared, as he did in an 1856 article about Borneo, that a trip to the East Indies would be worthwhile for one reason only: to taste a durian, the remarkably thorny tropical fruit with a horrible smell and a delicious, custard taste.

Wallace's vivid, sensuous descriptions were beyond the imagination of a Quaker teenager living in a plain stone house in the middle of a prairie. Hearing Wallace's tales about the South Pacific was a pivotal moment in David Fairchild's life. He suddenly realized that botany didn't have to be about familiar brown farm crops. Studying plants could lead to romantic adventures and exciting glimpses of life in exotic lands. Fairchild immediately developed a deep yearning to escape the farmlands of Kansas.

"It was this meeting with the great naturalist which started my longing to see, when I grew up, those islands of the Great East," he recalled.[9]

*   *   *

By the end of 1888, David Fairchild was ready to set off on the adventures that lay before him—and the world was ready for him. It was the birth of globalization, the gradual interweaving of people, cultures, and customs that has since transformed the American experience. Fairchild's romantic notions about exotic places began by the fireplace in his parents' home and grew stronger during college. His Puritan/Quaker parents had instilled in their youngest son a toughness that made him push on toward adventures in the world beyond Kansas. David Fairchild's romantic notions and single-minded determination combined to help transform American life.

*   *   *

After he finished college, Fairchild left the drab wheat fields of Kansas. At first he got only as far as Ames, Iowa, where he lived with his uncle Byron Halsted, who was a professor at Iowa State Agricultural College. Fairchild loved studying plants with Halsted. "I learned from him that investigation should be a series of exciting experiences," he remembered.[10]

Shortly after Fairchild arrived in Iowa, Halsted accepted a better offer from Rutgers University to run its new plant experiment station in New Brunswick, New Jersey. When Halsted moved east, Fairchild tagged along.

Even after the plague of locusts subsided in the 1880s, farmers throughout America were struggling to fend off other insects and diseases—two botanically separate but related afflictions—that threatened their crops. Because only a handful of American botanists were studying the problem, the field of plant pathology consisted of a small, clubby group of scientists. Halsted was an early and influential member; another was Beverly T. Galloway, a young scientist from Missouri who worked for the United States Department of Agriculture in Washington, D.C. Like Halsted, Galloway wanted to do scientific research that was useful to America's struggling farmers.

FIGURE 1. David Fairchild as a young man, before he began a lifetime of plant exploring. By permission of the Fairchild Tropical Botanic Garden, 0188, David Fairchild collection.

In the U.S. government's drive to understand insects better, the agriculture agency had had an entomologist on its staff since 1863, but in 1888 the government created a separate section to fight plant diseases. Galloway was chief of this division, a job that required sharp scientific thinking and strong practical skills. After he settled in Washington, Galloway started hiring smart, hard-working assistants. By the summer of 1889 he had found three young men and was looking for a fourth.

One July day Galloway took the train to New Brunswick to meet Halsted's smart young nephew from Kansas. The two men liked each other right away. Fairchild described Galloway as a "long-legged, thin man, with kindly brown eyes behind big spectacles . . . with a big smile that won him friends everywhere."[11] Galloway saw Fairchild as "a good, clean, lovable, uncomplaining boy."[12]

Before the afternoon ended Galloway offered Fairchild one thousand dollars a year to join his team of investigators. Fairchild accepted without much hesitation and within a few days, after abandoning his postgraduate studies, he was on a train to Washington, reading up on diseases in onions to prepare for his new job. The opportunity was, he said, "a bombshell."[13]

*   *   *

Fairchild arrived in the nation's capital on July 25, 1889, four months after the inauguration of Benjamin Harrison, a Republican from Indiana. The United States totaled thirty-eight, although four new ones—Washington, North Dakota, South Dakota, and Montana—would be added in November 1889. The country's population was a little more than 50 million. Farming was an enormously important segment of the economy: the market value of agricultural products was more than $500 million (more than $12.5 billion in current dollars). Young scientists working to improve agriculture were as valuable to the nation as rocket scientists would be seventy-five years later. The department had already hired Fairchild's friend Charles Marlatt as an assistant in the entomology division under Charles Valentine Riley, an internationally celebrated scientist who had written many articles to help farmers understand the locust plague.

Despite the national importance of farming, the U.S. Department of Agriculture had become a cabinet-level agency—one of seven—only a few months earlier. For decades, presidents had considered creating a separate office to help farmers, but many legislators, especially southerners, vehemently opposed granting the federal government any official role in the family farm, a fiercely independent American institution.

Congress had finally established the office in 1862 only because the southern states had seceded, leaving northern senators and representatives free to approve the legislation without opposition. To run the new agency, President Abraham Lincoln appointed a New Jersey farmer named Isaac Newton, a man some called the White House milkman because he supplied fresh butter to the president, an important qualification for the job. Newton started with only nine employees, but the agency's budget and authority expanded quickly as farmers clamored for help combatting the forces of nature.

*   *   *

By the summer of 1889, when Fairchild arrived in Washington, five hundred people worked for the U.S. Department of Agriculture and all were scrambling to meet their increased responsibilities. Scientific understanding of botany and entomology was poor at the turn of the nineteenth century; for example, pear blight, a disease that kills young branches and eventually whole trees, was the only bacterial disease of plants recognized at the time. Galloway's crew was a tight-knit group of young scientists who labored intensely to crack the mysteries hidden in plants.

Working conditions were uncomfortable. At first, the Section of Vegetable Pathology consisted of four scientists and two clerical workers. They were crammed into an old brick building near Pennsylvania Avenue with offices in a couple of small rooms four flights up in the attic. Their laboratory was squeezed into a damp, badly ventilated space in the basement. Workers risked death there from dangerous fumes and gas explosions. Nonetheless, Fairchild loved his job. "I immediately felt at home," he remembered, "for I realized that these men were pioneers in the study of plant diseases."[14]

Fairchild spent his days squinting into a microscope, trying to comprehend maladies that threatened apples and other nursery plants in the Northeast. Despite the narrow limits of his work, he believed he was having the adventures he had dreamed about in Kansas. "An atmosphere of scientific Bohemianism prevailed among us," he wrote later.[15]

Although he enjoyed the work, Fairchild was sometimes lonely living in a big strange city. One day he had become so sad that Galloway took him home for dinner and tried to cheer him up by playing the guitar and singing "Love's Old Sweet Song." Galloway's kindness was not enough, however; Fairchild wanted Walter Swingle, his best friend from Kansas, to join him.

After Swingle graduated from college in 1890, Fairchild pleaded with him to move to Washington, too. "You know very well what a hold Dr. Riley's division has upon the entomologists, do you not?" he told his friend on March 15, 1891. "There is no reason why if you can get enough pushing fellows together we cannot make this division as great and in my mind a greater means of doing work for the science."

Fairchild held out the remote but romantic possibility that the government would send them on out-of-town assignments. He reported that the department had already paid for one employee to join a plant expedition through Death Valley and might send Swingle to, say, Florida, a remote,

FIGURE 2. The earliest known photo of the staff of the section of plant pathology of the U.S. Department of Agriculture. From left to right: front row: David Fairchild, Beverly T. Galloway, Walter Swingle; back row: Joseph James, Theodore Holm, Merton B. Waite, and P. Howard Dorsett. Three other staff members—Erwin F. Smith, Effie Southward, and Newton B. Pierce—were working in the field when the photo was taken. By permission of Special Collections, National Agricultural Library.

mysterious outpost. These were opportunities, Fairchild promised in the letter, "which viewed from your standpoint may seem improbable but believe me to my mind are sure to come."

He pleaded with his friend to take the job. "If you will only come, Swingle, I'll share lots of things with you and try to make matters as pleasant for you as I can," Fairchild promised.

Swingle agreed, but because, like Fairchild, he was only nineteen, he needed written permission from his parents to become a government scientist. He arrived in Washington in May 1891, bringing his own microscope and botany books in several languages. As Fairchild had predicted, Galloway sent Swingle to Florida to figure out what was wrong with orange trees owned by the wife of an important New York politician.

Galloway also assigned fieldwork to Fairchild. First he went north to upstate Geneva, New York, to help commercial growers cope with diseases that were killing grapes, currants, and pears. Then he traveled to Chicago in 1893 to manage the U.S. Department of Agriculture's display of plant diseases at the Columbian Exposition. Charles Marlatt was in Chicago, too, managing a related exhibit of dangerous plant pests that featured 602 varieties of insects.

Running an exhibit at the World's Fair was arduous work even for young men. Fairchild, twenty-four, spent all day showing visitors wax models of sick, deformed vegetables and fruits and trying to explain complicated theories about plant pathology. He set up an exhibit of tiny pear trees that he dramatically infected with pear blight several times a day. "They would die before the very eyes of the visitors to the Exposition," Fairchild wrote.[16]

Whenever he had a few hours off from work, he tried to see everything else at the fair, which he later called "the greatest dream-city man had ever built."[17] One popular exhibit must have made a huge impact on Fairchild after the thrill of hearing Alfred Russel Wallace describe the South Seas. It was Java Village, an authentic, reconstructed community inhabited by 125 natives and one adult orangutan imported to the shore of Lake Michigan. The exhibit was built of bamboo, an exotic wood, and enclosed by a 10-foot-tall fence. Admission was 50 cents.

The world's fair assignment exhausted Fairchild and triggered the first of many physical and emotional collapses that he suffered throughout his life. The hard work and the brutal summer weather in Chicago "nearly broke my health," he explained later, "depleting my strength and enthusiasm to a degree which I had not dreamed possible."[18]

At the end of the Chicago assignment, he decided to quit his job at the agriculture department and find a better way to see the world.

Realizing his dream—to visit the real Java—was impossible, he believed. At the time, it was the most expensive destination in the world for Americans. The steamship ticket alone cost $500 (almost $13,000 in today's dollars), an amount that equaled half Fairchild's annual salary. Because Americans' opportunities to travel were expanding, however, Fairchild was able to devise an alternative that was interesting but not as exotic. He won a scholarship from the Smithsonian Institution to study at the Zoological Station in Naples, a marine biology center for visiting foreign scientists. It was one of the first times the station had accepted a scholar from America.

With much haste and with no obvious regret—he failed to finish an article on new fungicides before he left—Fairchild resigned from the government, despite older employees' warnings that he was making a big mistake. "They told me of young men who had gone to Europe and returned filled with newfangled notions which were of little use in practical American life," he wrote.[19]

This time the older men were wrong.

*   *   *

Fairchild's adventures began, as many significant things do, with a casual encounter. As he traveled to Italy in November 1893 on the SS *Fulda*, a 7,000-ton German ship on its first voyage from New York through the Mediterranean, he met Raphael Pumpelly, an American geologist taking his family to live in Italy because he couldn't find work in America. Fairchild confided to Pumpelly that he would have preferred to be going to Java but he couldn't afford to get there. Pumpelly knew another passenger on the ship who had visited Java and thought Fairchild would be interested in hearing his tales of the South Seas. One evening, when the ship was cruising past the Azores, Pumpelly brought Fairchild to the first-class cabins' smoking room to introduce them.

Pumpelly's friend was Barbour Lathrop, a forty-six-year-old millionaire from Chicago who was on his eighteenth—or maybe sixteenth or nineteenth, he had already lost count—voyage around the world. Elegantly dressed for dinner, Lathrop was relaxing when Fairchild and Pumpelly entered the room; he held a novel in one hand and a cigarette with a Turkish cherry-wood holder in the other. The stranger's pose was striking, and as Lathrop glanced up from his book, his appearance astonished Fairchild.

"As he twirled his gray mustache nervously and looked at me with those steel gray eyes of his, I thought he seemed the handsomest and the most severe and yet the most intelligent person I had yet met in my life," Fairchild remembered.[20]

Their conversation was awkward. Lathrop boasted that he knew Java well and pronounced that the view from the Hotel Bellevue in Java's capital city was one of the three most beautiful sights in the world. Lathrop described an unfamiliar fruit—something called a mango—he had eaten in Java. He bragged that he had hunted rhinoceros there and had narrowly survived the dreaded tropical fever.

As Lathrop talked, Fairchild said, he engaged in an elaborate personal ritual. "He would pause, blow the stub of his cigarette out of its Turkish holder, take a fresh bit of cotton from his case, tamp it down into the holder with his pencil and, with the precision which comes from long habit, fit a fresh Egyptian cigarette into the holder and light it," Fairchild remembered. "This performance was as much a part of him as his walk or voice."[21]

Fairchild had little to contribute to the conversation, so Lathrop dismissed him after several minutes. "I left quite awed, feeling that I had met one of the most widely traveled men in the world," Fairchild wrote later.[22]

*   *   *

They didn't speak to each other again on the ship, but Fairchild watched Lathrop from a distance during the rest of the voyage. He was impressed by what he saw. Although the sea was rough, Lathrop appeared to be the only passenger who didn't get sick. One night Fairchild, who was too poor to own the formal clothes required to eat in the ship's fancy salon, hid behind a pillar while he watched Lathrop preside as toastmaster at a dance hosted by the captain. "I never dreamed of seeing him again," Fairchild wrote later. "I don't recall that he said goodbye or showed me any courtesy in leaving."[23]

When the *Fulda* docked at Gibraltar, Lathrop left the ship to watch the French Foreign Legion in action in Morocco. Fairchild stayed onboard until Naples, where he began his own much less romantic project: studying algae from the bottom of the Bay of Naples.

They were two men leading different lives who happened to meet on an ocean voyage. Yet their casual encounter triggered a string of fateful decisions that transformed American food forever.

Good Knight of the Four Winds

Barbour Lathrop knew nothing about plants.

"I doubt if he could have told a maple from an oak," Fairchild remarked. Lathrop had probably never visited a working farm or read a botany textbook. He had, however, learned a lot about food during his journeys around the world. He loved to eat, a pastime that had taught him an important lesson: dining tables in other countries were filled with delicious dishes and ingredients unknown in the United States. The more meals Lathrop enjoyed, the more convinced he became that these foreign foods should be introduced to America. Lathrop had, Fairchild wrote later, "a deep-seated conviction which grew with his travels: that his countrymen should know of the foods of other nations."[1]

He was born on January 26, 1847, in Alexandria, Virginia. His given names were Thomas Barbour, but he discarded the first one early on and went by his middle name, which came from his mother, Mariana. She had grown up in the Piedmont region of Virginia before the Civil War in a family of prominent politicians. One uncle on her mother's side was James Barbour, who had been Virginia's governor before she was born. Another was Philip Pendleton Barbour, a Speaker of the U.S. House of Representatives and an associate justice of the U.S. Supreme Court. Both uncles were important men who lived near Mariana in majestic homes. Although her father, Daniel Bryan, had less impressive family connections than her mother did—he was a lawyer, poet, and a champion of struggling artists— Mariana Bryan was raised as a privileged southern belle. Everyone called her Minna.

In 1843 she married Jedediah Hyde Lathrop, an ambitious young man who had moved south from his birthplace in Lebanon, New Hampshire, to make his fortune in the nation's capital. Family histories say banking and investments made him rich, especially from his holdings in the institution that became Riggs National Bank. He abhorred slavery and considered himself a friend of President Abraham Lincoln. In addition to Barbour, the couple had a second son, Bryan, and two daughters, Minna, who died when she was a young woman, and Florence.

Barbour Lathrop learned early about the thrill of travel. He was four-teen when the Civil War began, and to honor the romantic traditions of his Virginia heritage, he enlisted in the Confederate army as a drummer boy. His military career ended after only three days when his parents tracked him down and dragged him home.

As a Union man, his father received so many serious death threats after Virginia seceded that the family was forced to flee Alexandria. Barbour Lathrop's family was, as a relative said, "torn from us by the intolerance of despotism."[2]

Barbour Lathrop remembered that none of his mother's friends came to say goodbye on the day they left, but Minna Lathrop thought she saw one hand flutter farewell through a closed shutter as they drove out of town in their horse-drawn carriage. The family was on the run for more than a year; they stayed first in Baltimore, then West Chester, Pennsylvania, and Yonkers, New York, finding temporary shelter in rooming houses along the way. The experience terrified Barbour's mother.

"I see nothing but anarchy and gloom before us and would not be surprised at anything," she told her parents in a letter from Yonkers.[3]

Despite the danger—or perhaps because of it—their desperate escape gave young Barbour a taste for exciting travel and a deep admiration for soldiers' courage and derring-do.

The Lathrop family wound up in Elmhurst, Illinois, a new suburb of Chicago, then a growing metropolis where Minna's brother, Thomas Barbour Bryan, was amassing his own fortune. Jedediah Lathrop built a large home in a neighborhood developed by Bryan. There the Lathrops were safe from southern sympathizers, but Barbour Lathrop's independent streak remained strong. Shortly after the family settled in Illinois, Jedediah Lathrop sent young Barbour away to boarding school in New York.

He was the only southern boy in the place, and the other students ostracized him because of his Confederate sympathies. He claimed that students neither looked at him nor spoke to him for two years.

When Lathrop was about sixteen, he graduated from boarding school and went to study at the University of Bonn. For two years he enjoyed the life of a smart, curious, and rich young American in Europe. He made many friends there, singing drinking songs with them in German beer halls and getting spruced up for nights out in Paris. "A beautiful youth in evening dress, he made his way like a young lord up the great mirrored staircases of the Opera, crowded with jeweled magnificent ladies and gentlemen," Marjory Douglas, his friend and unofficial biographer, wrote after his death.[4]

Lathrop befriended—quite innocently, he insisted—a Parisian woman who was a notorious courtesan and acquired social graces that remained with him for the rest of his life. "He gained sophistication, a polish of manner that could be charming or insolent as he chose and an undefeated self-assurance that would carry him anywhere he decided to go," Douglas wrote.[5]

While he was traveling through Europe, Lathrop met Giuseppe Garibaldi in Italy. Even the important Italian leader noticed the seventeen-year-old's independence. "Extraordinary people, these Americans," Lathrop said Garibaldi told him. "They let their children travel around in Europe without any attendants, alone."[6]

After a few years of wandering, Lathrop returned to America to attend Harvard Law School. He graduated in 1869 when he was twenty-two and moved to Chicago to work for Wirt Dexter, who handled big business and divorce cases. The job didn't last long, however, because Lathrop scorned the work and turned down Dexter's invitation to become his partner.

"A lawyer cannot tell the truth," Lathrop said later to explain why he rejected this lucrative offer. "He cannot express what he thinks about a case. I wouldn't be a lawyer for all the wealth in the world."[7] Lathrop always relished saying what he thought, even if his remarks were rude or insulting.

His parents probably hoped that after law school he would settle down in Chicago and live a comfortable life near his relatives, a family by then firmly established as civic leaders. But it was too late. By the time he was in his mid-twenties, Barbour Lathrop had become an insatiable traveler.

His father was so angry at his refusal to stay in one place that he cut off his allowance. In about 1870 Barbour Lathrop, well educated but broke, left Chicago and began to roam.

First he headed for New York City and got work writing for *Ironmonger's Gazette* and *Haberdasher's Review*, two long-buried journals. Next he visited Japan, where he met Horace Capron, a former U.S. commissioner of agriculture on a mission to help Japan's government introduce western agricultural techniques to the country. Somewhere along the way Lathrop worked as a surgeon's assistant on a ship and held a series of other long-forgotten jobs to finance his wanderlust.

"I paid every cent of my own expenses and earned money by hard work," he told a friend later. "As soon as enough was hoarded for a trip, I resigned whatever position I held and traveled."[8]

In the mid-1870s Lathrop moved to San Francisco to become a newspaper reporter, but getting assignments was hard. At first, he told Douglas, he was so broke that he stole food: early each morning he took bottles of milk from doorways and fresh rolls from bakers' carts. Eventually, after he had sold enough freelance articles to save a few dollars, he tracked down the victims of his crimes and gallantly reimbursed them.

Even when he was broke, Lathrop exhibited great style. "He wore always the finest clothing, despite a small salary, sported a silk hat by day and beautifully cut tails by night, always wore gloves and carried a walking stick," Douglas wrote.[9]

Lathrop eventually persuaded the editors of San Francisco's *Morning Call* and *Evening Bulletin* to give him regular assignments. One paper sent him to Sacramento to cover the California legislature. Another time he scooped the competition by scoring a firsthand account of the 1875 sinking of the SS *Pacific*, a ship laden with gold that went down off the coast of Washington state. Lathrop's assignments were sparse until he finally got his big break: in 1876 the *Bulletin* hired him not simply as a newspaper reporter but as a war correspondent.

The opportunity came when the U.S. Army launched a campaign to drive Indian tribes from the Sioux reservation on the northern Great Plains to clear the land for gold miners. For months, the Army pressured the Indians to relocate with a string of skirmishes and raids on Sioux villages. On

June 25, 1876, near the Little Big Horn River in what was then Wyoming Territory, the Indians pushed back, hard.

Sioux chief Sitting Bull and his men killed General George Armstrong Custer and his 306 troops in a bloody battle that a U.S. government official summed up as "utter annihilation." The only survivor was Comanche, a horse owned by one of Custer's officers. Also killed was Mark Kellogg, a reporter for the *Bismarck Tribune* in North Dakota and a stringer for the Associated Press. A few weeks after the Battle of Little Big Horn, the body of another reporter, a man sent out west from Springfield, Massachusetts, by the *Republican*, was found scalped after another skirmish with Sioux.

When top Army officers vowed to take revenge, the *Bulletin* hired Lathrop to chronicle the battle. Covering the Indian wars was a dangerous assignment, but it suited Barbour Lathrop's lifelong notions about romance and glory. Lathrop, twenty-nine, had admired soldiers since his brief stint as a Confederate drummer boy.

His assignment was to follow Brigadier General George Crook, one of the campaign's two commanders. Lathrop rushed from San Francisco to Fort Fetterman, an outpost on the edge of Indian Territory in Wyoming, but he arrived just after Crook and his troops had left. Lathrop was stranded at the fort with another reporter, Cuthbert Mills of the *New York Times*. Nonetheless, Lathrop was confident he would soon be covering a great battle.

"Stirring events may be looked for 'ere long," he wrote in a story the *Bulletin* published on July 17, 1876.

After an anxious week away from the action, the two civilians were thrilled at last to see reinforcements arrive at Fort Fetterman on their way to meet Crook.

"A pretty sight it was from the commanding hill on which the fort stands to see the regiment filing up the road," Mills reported.

He added an observation that was undoubtedly from Lathrop: "A gentleman who stood watching the regiment pass told me he had seen English, French, Prussian, and Russian cavalry in the field and, though the soldiers of those nations exceeded ours in neatness and uniformity of appearance of dress, yet they none of them presented such a thorough fighting look as the Fifth Regiment."

The reinforcements included General Eugene A. Carr and his favorite scout, William F. Cody, the famous Indian fighter. Cody had already begun a second career as an entertainer, but he had cut short the tour of his company, Buffalo Bill's Wild West Show, to join the campaign to kill Sioux after the Battle of Little Big Horn.

The troops and reporters left Fort Fetterman on July 26 and tramped for more than a week before they caught up with Crook. The days were uneventful. "The march from Fort Fetterman was over a disgustingly bare country, which seems to produce nothing but sage brush, alkali water, grasshoppers, and mountain fever," Lathrop wrote in an August 9 report.

He had time to make friends with a few officers, especially General Carr, a cultured man and Civil War hero who usually avoided reporters. Barbour Lathrop unleashed so much charm that Carr let him share his tent and write his stories in comfort at his portable camp desk.

On August 8, during a break in the long days of difficult but unexciting marching, the soldiers had a chance to relax. A friend of Cody's arrived on horseback and presented the scout with a bottle of whiskey, a rare and probably banned treat on the battlefield. Cody invited Carr to share the gift and, as the two men snuck off to drink it alone, Lathrop followed them.

"He, having the true nose of a reporter, smelt the whiskey from afar off and had come to 'interview' it," Cody wrote in one of his three autobiographies. "He was a good fellow withal and we were glad to have him join us."[10]

Despite the cocktail break, hunting the enemy was a terrible ordeal. The soldiers campaigned under great strain and confusion; in one day three thousand men crossed a single river thirteen times. In mid-August torrential rain drenched the troops. After almost a month of fruitless effort, soldiers were sick and desperate for food. And, worse, they hadn't avenged the massacre.

"No Indians yet and no prospect of finding any," Lathrop wrote in an article published September 5 after the troops had become miserable and angry. "A grumbling fever has become epidemic."

Crook made his troops keep looking—although later evidence showed that the Indians had scattered—but the civilians chose to leave early. Cody quit, saying he had more pressing work to do in show business back east. Lathrop returned to San Francisco at the same time.

On December 5, 1876, shortly after he arrived in the city, Lathrop's colleagues marked their respect for his reporting skills by inviting him to join the Bohemian Club. The club, which would later become a controversial secret society of rich and powerful men, was only three years old. It had been formed by a group of reporters who ate breakfast every Sunday at the home of James F. Bowman, an editorial writer for the *San Francisco Chronicle*. Breakfast grew into full-day sessions, with wine, piano playing, and lively discussions about journalism. Eventually the crowd grew too large and animated to continue to hang out in Bowman's home, and the club moved into its own building in downtown San Francisco.

At first the organizers allowed only reporters and editors to join and decreed that other bohemians—artists, actors, poets—could be mere honorary members. Newspaper owners were banned altogether.

The 1870s were a thrilling time to be a writer in San Francisco. Many early club members became famous for their intellect and wit. Henry George, the founder and editor of the *San Francisco Daily Post*, was a trustee. Ambrose Bierce was secretary. The group made Samuel Clemens a member, although he had quit his job at the *Morning Call* years before and moved to Hartford, Connecticut. Lathrop was twenty-nine when he joined.

The club's official annals described him in terms that suited Barbour Lathrop for the rest of his life: "Of a social disposition, traveled and well-informed, he had plenty to talk about and he talked; at the same time, he always talked well."[11]

Because he never stopped talking, Lathrop told Douglas that he became known as "the man with the iron jaw." Old-timers at the club said the only member who talked more than Barbour Lathrop was Rudyard Kipling.

By 1890 several members, who believed that the club had forsaken its bohemian roots and become too respectable, resigned to establish a rival club called the Pandemonium. Always loyal, Lathrop stayed a Bohemian. His room on the top floor of its San Francisco clubhouse remained his only home for forty-nine years.

* * *

During his first trip to Japan in the 1870s, Lathrop had learned about the Ainu, a mysterious tribe of Japanese aborigines. The race was thousands of years old, but by the 1870s civilization threatened its survival. One remarkable custom was that each Ainu man traditionally grew his hair, beard, and mustache so long that they created a curtain over his face that he parted with his hands before eating. The tribe was particularly intriguing to westerners because Japan was one of the most mysterious lands on earth: the emperor barred virtually all foreigners from traveling beyond Yokohama, the major port.

Lathrop traveled inside Japan anyway and, using many ruses of an experienced reporter, maneuvered his way to the northern islands of Japan where the Ainu lived. He gained the tribe's trust by brandishing his set of false teeth and setting paper on fire with a magnifying glass. He took extensive notes, intending to write a book about the culture. He never finished it, but in true Lathrop style he had a colorful excuse.

He was staying in a hotel in Callao, a large port in Peru, when a fire broke out in the building. Lathrop had to choose between saving his Ainu material and rescuing a woman trapped in her room. She was terrified, screaming, and naked. He chose the woman. As they fled the burning building, Lathrop wrapped his coat around her and announced, "Madam, you cannot appear that way even in these circumstances."[12]

The fire destroyed all his notes and his unfinished manuscript. Lathrop never again attempted to write another book, even his autobiography.

*   *   *

While Barbour Lathrop was busy knocking around the world, having adventures, and collecting colorful anecdotes, his family's fortunes had flourished in Chicago. After the Civil War ended, his uncle Thomas Barbour Bryan built Graceland Cemetery, a significant urban development that was the city's first landscaped burial ground. He hired his nephew, Bryan Lathrop, to manage the cemetery, a job he apparently did well. Creating Graceland would probably have remained the family's biggest accomplishment if not for the Great Chicago Fire of Sunday, October 8, 1871, a day that created one of the biggest real estate investment opportunities in American history. The fire triggered a chain of events that transformed urban architecture and, in the process, produced the personal fortune that bankrolled America's first plant expeditions.

The fire roared through downtown Chicago for three days. It killed more than 200 people, left about 100,000 others homeless, and burned down 18,000 buildings on more than 2,000 acres. Eighty blocks, about a third of the city, were reduced to rubble. Chicago was devastated.

Four years after the disaster, as the city struggled to rebuild, Barbour Lathrop's brother Bryan married Helen Aldis, the daughter of a Washington, D.C., lawyer. After the wedding, Helen's younger brother, Owen Franklin Aldis, a twenty-two-year-old lawyer, moved to Chicago to work in the real estate business with his new brother-in-law.

Aldis opened his office in what was called the Portland Block, a four-story structure that was one of the first buildings erected after the fire. In February 1879, when Aldis was twenty-six and struggling to support a new wife, he received a letter from Peter Chardon Brooks in Boston. Brooks was an investor who wanted to make money rebuilding Chicago. Brooks hired Aldis to find promising properties for him and his brother, Shepherd.

Aldis didn't look far for their first deal. He bought out of foreclosure the Portland Block, the building that housed his own office. Brooks was satisfied by Aldis's quick work and asked for more. He told his young agent that he was especially interested in property where he could build structures that rose higher than four floors. "Tall buildings will pay well in Chicago hereafter, and sooner or later a way will be found to erect them," Peter Brooks wrote Aldis on March 21, 1881.

For $150,000 cash—the Brooks brothers shunned mortgages—Aldis bought a parcel nearby on which they built the Montauk building, a ten-story structure that became the world's first skyscraper. Later he defied conventional wisdom by buying land on West Jackson Boulevard for the Monadnock, a seventeen-story building that was then the largest commercial structure in the world. It is now in the center of Chicago's business district.

"Everyone thought Mr. Aldis was insane to build way out there on the ragged edge of the city," Edward A. Renwick, a Chicago architect, said later.[13] His role was so important that Louis Sullivan, the world-famous architect, asserted that Aldis, along with the inventor of the elevator, were the two men most responsible for creating the modern office building.

The daring team of Aldis and Brooks went on to develop at least a dozen more skyscrapers in downtown Chicago, each a few floors higher than the last. Records show that Aldis financed these projects by creating real estate investment trusts, then an unknown financial instrument. These were tight, complex networks of relatives, mostly brothers and brothers-in-law. Bryan Lathrop was included in many deals, and because he handled his own family's investments, he often shared the opportunities with his parents, sister, and brother. They all got rich.

As important players in Chicago's social and cultural life, the Lathrops had expensive lifestyles to maintain. In 1879 Florence Lathrop had married Henry Field, the brother and business partner of the merchant Marshall Field; she was twenty-one and he was thirty-eight. They had two daughters and lived happily together until Henry died in 1890. As a tribute to him, in 1893 she acquired the two huge bronze lions that still flank the entrance to the Art Institute of Chicago. The same year she married Thomas Nelson Page, a writer and diplomat, and became a prominent hostess in Rome, Washington, and York Harbor, Maine.

By 1880 Lathrop's father and mother were wealthy enough from their investments to describe their occupations to a federal census taker as a "gent" and "lady of leisure." Jedediah Lathrop died in 1889 and, his longtime feud with his younger son apparently forgotten, left money to Barbour Lathrop.

With his father's fortune and his brother's investment guidance, Lathrop was financially secure. As one of many interconnected brothers-in-law, he owned interests and a share of the rents in remarkable office buildings in downtown Chicago like the Caxton, the Old Colony, and the Marquette.

Owen Aldis and Bryan Lathrop had not only transformed American architecture but, in those days before a federal income tax, they had given Barbour Lathrop more money than he knew what to do with.

Barbour Lathrop was forty-two when his father died. His inheritance allowed him to stop hustling for newspaper assignments and officially change his occupation to "gentleman," according to his passport application. Marjory Douglas said sudden wealth didn't alter his way of life, however. "He did not want to get married or live more elaborately," she wrote. "All he wanted to do now was travel."[14]

He loved most being at sea. He craved "the rush and surge and burst of white spray over the bows of innumerable ships, cutting through the roaring blue waves of all the oceans or sliding through the quiet, green glassy waters of innumerable strange warm seas and straits, among green, mysterious unknown islands," Douglas wrote.[15]

Lathrop also enjoyed showing off his sudden prosperity and insisted on being treated like an important man. He gave away money constantly, usually to individuals, not institutions, probably because he wanted full credit for his largesse. One tale of his grand gestures was that whenever he met a woman who was going to Paris for the first time, he gave her $500 to spend on herself, a gift worth almost $13,000 in today's dollars.

He told Douglas that he once broke the bank in Monte Carlo and, instead of keeping his winnings, spent all the money on a lavish banquet for the guests at the casino the night of his triumph.

His travels had no purpose except socializing and sightseeing. Eugene Field, his friend and fellow Bohemian Club member, fondly labeled him "Sir Barbour, the good knight of the four winds." He traveled to tell stories about traveling.

"He was an unforgettable character," Douglas wrote in a *Reader's Digest* article about Lathrop, "often inescapably the hero of his own stories by his dash, his ingenuity, his ready wit, his constant courage."

After a few years, however, Lathrop began to grow weary of aimless sightseeing and devised a plan to occupy himself. After he discovered unfamiliar foods and plants in markets and restaurants around the world, he vowed that, because no one else was bringing samples of them to America, he would do it himself.

At first he was mostly interested in medicinal plants. When he learned about the coco plant's narcotic powers in Peru in the 1880s, for example, he sent leaves to the Academy of Sciences in San Francisco for analysis. Unfortunately, no one there did anything with the gift. Lathrop, furious about the slight, realized that he needed a better scheme.

He wondered if foreign food plants could be introduced to America to help farmers diversify their crops. He understood that he couldn't handle such a big project on his own; he needed to link up with an expert who understood botany and agriculture to advise him on which plants would be useful in America.

He found one candidate, a young man whose name has been forgotten. The two left San Francisco on a voyage around the world to hunt for new fruits and vegetables. They had gotten only as far as Hawaii, however, when Lathrop's partner fell in love with a young woman. He pleaded to be excused from their mission so they could get married. Lathrop agreed.

"Lathrop bowed gracefully to a superior force and went on his travels, alone but disappointed," Douglas wrote.[16]

He remained single, rich, and alone, although at forty-six, no longer quite young. Toward the end of 1893 he booked passage on the SS *Fulda* for yet another purposeless voyage around the world. After his brief meeting with Fairchild in the smoking room, however, Lathrop's travels would never be the same.

3

The New Year's Resolution

After Fairchild and Lathrop separated on the *Fulda*, Fairchild arrived in Naples and immediately recognized how unexciting American meals had been. "No sooner had I landed in Italy that I began to get a perspective on the limited number of foods which the fare in my home and in American boarding houses had brought to my palate," he wrote later.[1]

His education began in a small restaurant where he usually ate lunch. There he sampled his first foreign food: a dried fig, a wickedly sweet morsel for a young man raised on boiled vegetables. He tried vermicelli with a sauce of tomatoes, a fruit whose possibly poisonous qualities were still being debated in America. He enjoyed Italian pasta so much—it was chewy and flavorful, not the mushy kind made with soft American wheat—that he collected fifty-two shapes and mailed them to friends in Washington.

When he wasn't eating, Fairchild was happily working under near-monastic conditions as a full-time scientific researcher at the Naples Zoological Station. "There was a feverish but quiet excitement about the place in those days when everyone stood, as he thought, on the brink of some discovery," he wrote later.[2] He spent his days peering intently through a microscope at slimy green seaweed from Naples Bay, trying to understand its cell structure.

One month after Fairchild began his studies, a laboratory assistant interrupted him and presented the heavy, engraved calling cards of Barbour Lathrop and Raphael Pumpelly. Fairchild was flummoxed. Was it possible that the *Fulda*'s most distinguished passengers had taken the time to find him? He cleaned up his small desk as fast as he could as the older, well-dressed men marched into the crowded laboratory.

"What are you doing, Fairchild?" Lathrop demanded.[3]

As he brushed off two chairs for his guests, Fairchild began to explain in detail about seaweed and cell division, but Lathrop interrupted him. Since their meeting on the ship, he had asked important friends in Washington about Fairchild and had been impressed by their reports. Now he wanted to help him realize his dream of visiting Java. "I've come to tell you that I've decided to invest a thousand dollars in you," Lathrop announced. "I think you are a good risk. I'm not interested in you personally, understand, it is just an investment."[4]

The pronouncement amazed Fairchild. No one had ever given him more than fifty dollars in his life. One thousand dollars had been his annual salary in Washington. "The bewilderment and elation of that day and the days following were such as I imagine the pauper must have felt when he found himself a prince," Fairchild wrote later.[5]

Fairchild realized, however, that despite his sound education, he was an inexperienced botanist who was not ready to take full advantage of the opportunities in Java, so he stalled. Fairchild told Lathrop he needed two years of preparation before he would be qualified to go to Java. Lathrop agreed to wait.

Four months later Fairchild abandoned the algae from the bay and left Naples to begin his own independent study program. He started in Rome where, as a young man of Puritan-Quaker stock, he was shocked by what he saw in museums. "The magnificent examples of art frequently depicted naked forms—things taboo in my childhood—and left me somewhat aghast," he recalled.[6]

From Italy he traveled briefly through Hungary and Germany, then the world's center of important scientific research. In all he spent a year and a half as an itinerant scholar, staying with friends from college and enrolling in various academic programs. He spent a semester in Breslau, Silesia; he passed the winter of 1894–95 in Berlin. In Munster in April 1895 he studied fungi—material at the forefront of early plant pathology—and tried to pass as an authentic German. There he slept at night with a contraption on his head that forced his mustache to curl up at the ends like that of young Kaiser Wilhelm.

Fairchild continued to pick unromantic-sounding subjects to study: his projects included three weeks on molds in horse manure and six months on fungus in frogs' intestines. In Bonn he roomed with Walter Swingle, who had left his own government job in Florida to study cell structure for a year. Fairchild's early peregrinations taught him some science, a smattering of languages, and a habit of recording, almost as notches on his belt, the names of all the exotic places he had visited—the longer and more unpronounceable, the better.

Between semesters at German universities, Fairchild arrived at another turning point. He received a letter from a former colleague at the agriculture department asking for a favor. A California farmer wanted to establish an orchard of citrons, the lumpy-skinned citrus fruit whose candied peel was an essential ingredient in Christmas puddings and wedding cakes. Since Americans were forced to import the fruit from Corsica, Fairchild understood citron cultivation could be a potentially lucrative new industry for American farmers. "Though no one person eats in a year any large amount of citron, yet everyone eats a little," he wrote later. Fairchild's colleague wanted enough cuttings from Corsica to start an orchard in Monrovia, California. It probably seemed like a simple task to the official in Washington, but Fairchild knew getting the cuttings would be a dangerous business. "I was nervous and had been advised that the Corsicans were not inclined to let scions of their fine citron trees go out of the country," he recalled.[7]

Energized by the challenge, he accepted the assignment anyway. "This was my first commission to work for my government in a foreign land," he bragged.

Fairchild, who assumed that the U.S. secretary of agriculture had approved the assignment, set off alone for Corsica in the summer of 1894. He spent his own limited money to stop in Switzerland and Germany on the way and endured a rough night at sea as he crossed from Italy to Corsica.

When he finally landed in Bastia he received a telegram from Washington with bad news. The secretary, J. Sterling Morton, a man who boasted to Congress every year about how much of his department's budget he hadn't spent, had not approved Fairchild's trip. The telegram read: "Secretary refuses authorization" because Morton saw no good reason to bring new plants to America. Fairchild was shocked, but he made the best of the situation. "There I was, with an adventure on my hands," he remembered, "and I enjoyed it."[8]

As Fairchild had expected, Corsica did not welcome him. He spent days ambling along dirt roads on a donkey, trying to locate citron orchards. He ate nothing but figs, his new favorite food. One day he passed the time by taking photographs with his clumsy camera, a model mounted on a tripod and covered with a black hood. This suspicious activity led immediately to his arrest. Fairchild protested, but he was unable to explain himself to the gendarme. He tried speaking French. He tried Italian. "I could neither understand him or make myself understood," he wrote later.[9]

After spending hours in a jail that "would rival in filthiness any that the Inquisition ever had,"[10] Fairchild discovered in his pocket a fifteen dollar check from the United States government engraved with former president Ulysses S. Grant's face. Waving the check, Fairchild persuaded the Corsican cop that the paper was a United States passport. The officer released him.

As he rushed away from Corsica after this close call with prison, Fairchild stole a few cuttings from citron trees along a road and hid them under his coat. Unequipped with material to protect the branches from drying out on the long voyage between Italy and America, he jammed the sticks into raw potatoes, packaged the lot and mailed them. The potatoes provided enough moisture to nourish the cuttings, which survived the trip to Washington. Officials sent the twigs to California, where they launched a profitable business.

David Fairchild had become a plant explorer, and his first discovery was a big success.

* * *

At the end of 1895, by which time Lathrop had traveled around the world twice more, Fairchild told Lathrop he was ready to go to Java. In April 1896 he bought an excellent microscope and left Genoa on a ship of the Netherlands-India Packet Boat line. The steamer crossed the Mediterranean to Port Said, traveled through the Red Sea to Aden, and finally entered the Indian Ocean, all exotic, exciting locales to add to Fairchild's growing list of ports of call.

The ship landed on the west coast of Sumatra at the village of Padang, a collection of low buildings strung along the waterfront and backed by thick jungle. Fairchild was finally in the South Seas, on the verge of seeing the world he had dreamed about in Kansas. He never forgot the thrill of his first visit.

"The memory of that first tropical night on shore and of the noise of the myriads of insects and the smell of the vegetation and the sensation of being close to wild jungles and wild people sometimes comes back to me even though millions of later experiences have left their traces on my brain," he remembered in his book *Exploring for Plants*.

The ship's second stop was Wallace's beloved island, the land of his dreams. "You approach Java with a feeling of how beautiful and lovable everything seems," Fairchild wrote later.[11]

Java, then a profitable Dutch colony, is about the size of New York State. A land rich in vegetation and lined from east to west with thirty-eight volcanoes, Java had been a Dutch colony since 1800 when Holland took possession of the island and other nearby territories from the Dutch East India Company.

The Visitors' Laboratory at the botanical garden in Buitenzorg, a city now called Bogor, was, like the Zoological States in Naples, an unusual spot where botanists from around the world worked together. This spirit of shared scientific inquiry among researchers of all nationalities and all specialties stayed with Fairchild for the rest of his life.

"The institution was to discover and bring to light a knowledge of the plant life of the tropical world," Fairchild wrote later. "Not for the uses of Holland and Netherlands India alone, but for the whole world of plants—a world which knows no national boundaries, a world which constitutes a vast, magnificent realm of living stuff destined to be of interest to the human race for all time."[12]

Fairchild picked a narrow, highly specialized topic to study: Do termites grow mushrooms in the tropics?

Fairchild immediately fell in love with Java's food and landscape. The view near the capital's major hotel was spectacular, just as Lathrop had said. It was a sight that many western visitors praised. "You ought to see the view from this hotel: the Salak volcano, a perfectly even cone with simply glorious vegetation, great bamboos, like waving ostrich feathers, a hundred feet long, palms of fifty sorts and flowering trees and orchids in great profusion," a friend of Fairchild wrote when he visited years later.[13]

Meals in Java offered one sensuous food after another. Fairchild enjoyed elaborate lunches of spicy rijstaafel (literally "rice table" in Dutch) with dozens of portions of chicken, fish, eggs, and other dishes, all seasoned with a range of curries and chutneys. "The first mouthful of this mixture brought perspiration to one's face," he recalled. "However, in the end, one experienced a sense of well-being and drowsiness which admirably suited the tropical custom of a siesta after luncheon."[14]

Most remarkable were the unfamiliar, even bizarre tropical fruits. It was in Java, in the summer of 1896, that David Fairchild began his lifelong love affair with one food: the mangosteen, a tropical fruit completely unknown in America. Fairchild first noticed it when he was seated on the veranda of his hotel in Buitenzorg. He saw "coming towards me a Javanese coolie in his turban and batik sarong with a long bunch of fruits to sell," Fairchild said later. "They were mangosteens cleverly tied with strips of bamboo to make a solid attractive package."[15]

The fruit's tough reddish-brown shell seemed impenetrable until Fairchild's hotel neighbor showed him how to get inside. "With quick turn of the hands, I could break open a mangosteen fruit without crushing the ivory white delicate mass inside and dexterously get that mass into my mouth," he wrote. "It was indescribably delicate and delicious. It melted in the mouth like the most delicate plum would and it reminded me of all sorts of flavors from pineapple to vanilla with a suggestion of something that I have never been able to find in my category of flavors."[16]

Four years later he launched a lifelong but ultimately unsuccessful push to cultivate them in America. His enthusiasm mirrored the fascination of Queen Victoria, who in 1855 allegedly promised to pay 100 pounds to the first person to bring her a single mangosteen.[17] She failed, too.

* * *

One afternoon Fairchild was shocked to receive a letter from Lathrop announcing he was on his way from Chicago to Java, accompanied by his brother Bryan, Bryan Lathrop's wife, Helen, and Carrie McCormick, a young Chicago heiress. They planned to spend about a week in Java and wanted Fairchild to show them around. Within days, the elegant foursome arrived, looking out of place in the jungle. "An atmosphere of dignity, culture and wealth surrounded them," Fairchild recalled.

At first the visitors made him uncomfortable. "I stood awkwardly answering Mr. Lathrop's questions, very ill at ease," he said. Fairchild believed his appearance shocked Lathrop. "I was emaciated after a touch of fever and, moreover, was dressed in the conventional two-piece white duck suit adopted by the Dutch, which buttoned up under the chin and was extremely unbecoming to everyone," he reported.[18]

After only a few days of sightseeing with the others, Lathrop announced he had spent too much time with the group and asked Fairchild to postpone his studies so he could explore Sumatra's west coast with him. "Fairy, there is nothing in those termites," Lathrop said, using one of the many nicknames he made up for his protégé. "Leave them alone and come along with me and I will show you the world."[19]

Persuading him was surprisingly easy because, Fairchild admitted later, he had already had enough of Java's tropical paradise. Fairchild prized the freedom to study whatever he liked, but the heat and the rawness of nature were too much for him.

"Life here lacks stimulus along [the] lines of conversation," he wrote to Swingle on September 22, 1896. "I wish there was more spiritual enjoyment. It is evident if a man stays here long he must become more or less of a beast."

Still, Fairchild hesitated to accept Lathrop's invitation because he thought he was too inexperienced to please his benefactor.

"He could not dance," Douglas explained. "He could not make small talk. He was shy with ladies. His clothes did not fit. He could not even tie his necktie properly."[20]

Yet Lathrop had already spent enough time with Fairchild to notice his conscientiousness and recognize he would make a good partner in his project of introducing foreign food and other useful plants to America.

As usual, Lathrop was curt in his praise. "The only encouragement David Fairchild received came when he was found working late over a report instead of taking part in all the talk and laughter in the smoking room," Douglas wrote. "Lathrop said, 'You're a worker, Fairchild, whatever else you aren't.'"[21]

FIGURE 3. David Fairchild and Barbour Lathrop relaxing in their stateroom while cruising off the coast of Sumatra on Christmas 1896. By permission of the Fairchild Tropical Botanic Garden, 9933, David Fairchild collection.

Lathrop was determined to recruit Fairchild to join his project. He invited him on a cruise of the South Seas so that he could see, taste, and smell the exciting collecting opportunities.

The tour began a few days before Christmas 1896 when their ship stopped on Sumatra early one morning. The view there was also thrilling. "As you steam into Emma harbor on Sumatra's west coast, your mind is overpowered by the sight of the verdure-covered volcanoes and trackless forests stretching away into the unknown and undiscovered," Fairchild remembered.[22]

After they landed, the two men toured the public market in a settlement called Pandang. It was a noisy, crowded place that offered a cornucopia of strange cultivated fruits and vegetables. Fairchild was immediately intrigued. The visit "showed me how many new and interesting food plants there were if only we had an established place where they could be sent," he wrote.[23]

Along the road out of town Fairchild saw fantastic wild plants he had never dreamed existed. He noticed yellow raspberries growing on the roadside. He shook a slim bamboo shoot, thirty feet tall, until it collapsed on his head. The sights were marvelous.

The decisive point came a few days later when they left Sumatra for Singapore. On December 31, 1896, the men were on board a 1,337-ton Dutch steamer called the *Maetsuycker*. They were relaxing after dinner in Lathrop's comfortable cabin and arguing about scientific inquiry. Fairchild began the discussion by praising pure science with its rigorous laboratory work and careful microscopic analysis. Yet his adventures had already changed him; he had been fascinated by the bounty of mysterious, delicious foods he had seen from Naples to Java.

Lathrop, perhaps sensing Fairchild's changing attitude, bore down on him, countering that actual experience was better than scientific study. Leave the laboratory, he insisted, and devote your time to practical work. Finally Lathrop unveiled his plan. Let's introduce these strange, foreign plants to America and see which ones take root, produce fruit, and make money for farmers and merchants. At the time, only about 2 percent of the world's edible plants were cultivated in America, and the typical farmer grew only about twenty of them.[24] Lathrop wanted Americans to open their mouths to new foods.

"He began to lay before me his idea of what a botanist could do if he were given the opportunity to travel and collect the native vegetables, fruits, drug plants, grains and all the other types of useful plants as yet unknown in America," Fairchild wrote later.[25]

It was a long evening of lively debate, and in the end, Lathrop won. Fairchild agreed to join his project. He would abandon his cloistered studies in Java and take up the mission of foreign plant introduction. As the clock approached midnight, David Fairchild promised Barbour Lathrop that he would spend his life searching the globe for new foods.

"Without Barbour Lathrop to goad him into an entirely different life work," Douglas wrote later, "to pay his salary and his expenses on their long wanderings, David Fairchild might have become a quiet, little-known if distinguished plant pathologist and entomologist, a scientist-scholar whose life might have been lived almost entirely within the walls of some laboratory."[26]

The Golden Age of Travel

Once before, a high-minded American had launched an ambitious plan to collect and grow foreign plants in the United States. That adventure ended in a bloody tragedy.

The earlier scheme started in Campeche, Mexico, in 1827 where Henry Perrine, a thirty-year-old doctor from Cranbury, New Jersey, was stationed as the American consul. In September U.S. Treasury Secretary Richard Rush asked Perrine and the other diplomats around the world to send promising seeds and plants to Washington for testing. It wasn't an unusual request; U.S. officials had often sought new food and other products to make the nation more self-sufficient. Rush's predecessor had issued a similar appeal in 1819; before that Thomas Jefferson and Benjamin Franklin had encouraged Americans living abroad to hunt down plants that might be valuable at home. "The greatest service which can be rendered any country is to add a useful plant to its culture," Jefferson wrote in 1800, a remark that later American plant explorers frequently quoted with pride.[1] Jefferson had followed his own advice: he once smuggled grains of rice from Italy to Virginia in his coat pocket even though Italian officials could have executed him if he had been caught.[2]

Like Jefferson, Perrine, who was a botanist as well as a physician, had great faith in useful plants. The only consul to answer Rush's letter, he sent seeds of many exotic plants to Washington. For a decade, Perrine identified and collected tropical and subtropical plants throughout Mexico that he hoped could be transplanted to the American South. One official said later that Perrine devoted "his head, heart and hands" to introducing tropical plants.[3]

In 1838, he persuaded Congress to take a chance on his conviction—some used the word *obsession*—that the United States had the right climate and soil to grow mangos, sisal, cinchona (for quinine), and other tropical plants. That year lawmakers granted him the right to establish a nursery in southern Florida, a region most people considered worthless for farming or almost anything else.

The grant was general, not specific, since the region was virtually uninhabited. Congress allowed Perrine to use whatever property he wanted within a 23,000-acre site on the empty mainland below Miami, which was then little more than a trading post. Before heading south, Perrine carefully planned his experiment garden and collected seeds of more than two hundred promising varieties.

When he was finally ready to start planting, however, the timing was terrible. Perrine and his family arrived on Christmas Day 1838 when native Seminoles and Florida's few white settlers were fighting a brutal war. To wait for the hostilities to end, the Perrines settled temporarily on Indian Key, a small island off the Florida coast.

Perrine rented a house from John Jacob Houseman, a New Yorker who ran the island and made a living by servicing—or, perhaps, stripping—ships wrecked on its shoreline. The Perrines and sixty-five other inhabitants believed that its distance from the mainland protected Indian Key. And it had been safe—until the summer of 1840 when Houseman taunted the enemy by offering to catch or kill every Indian in south Florida for a bounty of two hundred dollars each. The situation quickly grew tense.

At about 2 a.m. on August 7, 1840, sudden gunshots and war whoops woke the Perrine family—Henry and his wife, Ann, and their three children, Sarah, seventeen, Hester, sixteen, and Henry, thirteen. A few hours earlier Seminole Chief Chekika and two hundred angry warriors had boarded seventeen canoes and raced in the dark to Indian Key to avenge the island owner's threat. They attacked Perrine's big house first, giving Houseman and most of the others time to escape.

The basement of Perrine's house, which was built on the island's edge, was a cavity called a turtle crawl filled with water; the space was used to

FIGURE 4. Henry Perrine's house on Indian Key. Photo from the collection of Jerry Wilkinson, Historical Association of Southern Florida, by permission of History-Miami.

trap terrapin, then a popular dish in fancy restaurants up north. Ann Perrine and the children rushed into this basement, huddling together in the dark as water lapped at their necks, while Henry Perrine covered its trap door with the heavy trunk of seeds he hoped to plant in Florida. Then he went outside to plead with the Indians to leave him and his family unharmed. Perrine spoke Spanish to distinguish himself from Houseman, the New Yorker. "I am a physician and will go with you to heal your people," he told them. They understood him and left—temporarily.

The Indians soon returned, however, rushing into the house and chasing the doctor upstairs to a cupola on the roof. Perrine slammed the door behind him and strained to fend off his attackers. The family, shivering in the water below, heard everything that happened. "For a few moments after they swarmed up the stairs after him, there was a horrid silence, only broken by the blows of their tomahawks upon the door, then a crash, one wild strike, a rifle shot and all was still," his daughter Hester wrote later.

The Indians set fire to books in the family library to destroy the house, as Ann Perrine and the children sneaked away in the water to safety. Perrine's son returned a few days later to search for his father's body. He found only a thighbone, several ribs, and part of his skull. The fate of his seed chest is unknown.[4]

After the Perrine tragedy, no one suggested any other grandiose schemes to introduce foreign plants in America until Lathrop and Fairchild came along more than fifty years later. By then, most of the world, including south Florida, was at peace.

*   *   *

In 1897 the two men joined the growing number of Americans who had begun to travel to enlighten and amuse themselves, not to trade goods or make money. After the end of the Civil War, commercial steamship lines had opened old trading routes to tourists, offering luxurious cabins and European-style cuisine to the idle wealthy. When affluent sightseers arrived in port, they were welcomed at stately, modern hotels that were well staffed and comfortable. Travelers often brought six or seven steamer trunks with them so they could wear their most fashionable clothes while they visited the local sights. The end of the nineteenth century was the golden age of world travel.

When Fairchild and Lathrop began the adventures that would change America's eating habits, they looked like improbable companions. Lathrop was tall, slim, and always well dressed; in bearing he resembled the military men he admired. He carried a cane and wore a hat wherever he went. Fairchild, in contrast, was gawky and uncertain and rarely wore clothes appropriate to the occasion, whatever it was. Lathrop was demanding and critical; Fairchild was constantly frazzled.

In the beginning Lathrop, who had flashing dark blue eyes and expressive bushy eyebrows, called Fairchild "my investment," with a little bit of a sneer. Fairchild, fully aware of the contrast, felt inadequate. "Somehow I could not do anything quite to suit him," he admitted.[5]

Fairchild was so socially awkward that he agreed to one condition of working with Lathrop: he promised not to get married while he was exploring for plants.

*   *   *

Their expedition began immediately with a leisurely cruise to Singapore and Siam. They left the ship first in Bangkok, but they hadn't had time to arrange what Lathrop considered proper lodgings for each of them there. Lathrop took the only room available and made Fairchild sleep on the floor of his hotel's giant ballroom. The hard floor was uncomfortable, but the electric lights that illuminated the room were worse. Unable to sleep,

Fairchild unscrewed each light bulb to darken the room. He was astonished by what happened next.

"As I turned out the last light . . . I heard the rustling noise of something scuttling over the floor," he recalled. "It was a noise which I had not heard before, and I screwed the bulb back again to behold hundreds of the largest, blackest cockroaches I ever saw—thick, ugly brutes—running for cover under the baseboard of the ballroom."[6] Fairchild cowered under mosquito netting for the rest of the night and slept lightly.

He witnessed another unexpected sight a few days later when he and Lathrop attended a young Thai couple's wedding dinner. It was a special occasion because the Crown Prince of Siam also attended the feast. Fairchild found the food unfamiliar and the formal etiquette bizarre.

"During the thirteen-course dinner, every dish was strange to us except the rice," he wrote later. "Each course was noiselessly placed on the table by a servant deferentially crawling on his knees. Not a person stood or walked erect while the prince and his guests were at the table. At the close of the long meal, the wives appeared and even those of royal birth all hitched themselves across the floor like a child who has not yet learned to creep."[7]

As witnesses to the wedding ceremony, Fairchild and Lathrop were obliged by local custom to trickle perfumed water down the bride and groom's necks as the couple knelt together with their foreheads touching. "If the others poured as much water from the jeweled conch shell as I did," he wrote later, "the poor bride and groom must have been well soaked."[8]

Because one of Lathrop's many travel rules was never to stay long in any place, the pair had a dizzying itinerary. From Bangkok they returned to Singapore, then sailed to Ceylon and Australia. From Auckland, they took a seven-day voyage to Fiji where they met Thomas Hughes, an English sugar planter who spoke fluent Fijian. Hughes rented a boat and agreed to show them around the archipelago, which was then known as the Cannibal Islands. "We landed one morning on the island of one of the minor chiefs," Fairchild wrote later. "He was a cannibal, of course, as they all were in those days."

Hughes led them slowly to the chief's home, a large, thatched building raised on rocks from the ground. "Once inside, we found ourselves on finely woven mats of pandan leaves covering a layer of ferns so deep that it was difficult to walk," Fairchild remembered. "The chief, a tall, rather handsome man, was attired for our reception in a white dress shirt, a loin-cloth into which his shirttails were loosely tucked, and a black four-in-hand tie around his neck. Waving a long flyswatter switch made of the midribs of a coconut frond, he advanced slowly toward us."[9] The chief also wore eyeglasses without lenses.[10]

The sugar trader graciously introduced the men. "Mr. Hughes explained to him that the president of the United States and his private secretary had come to call," Fairchild wrote. "The chief was highly honored and invited us to sit down with him on the floor." The meeting was polite, and the conversation was mostly about food. The American visitors gave their host ice cubes from Hughes's boat, objects he had never before seen or tasted. The chief told them about his many cannibal feasts, a ritual that had recently been banned. "Although we might know about this hard, wet stone which seemed so hot," the chief told his guests, "we white men had never tasted delicious tidbits of barbecued human being."[11]

The chief complained that roast pig, the substitute he was forced to eat, was not as delicious as "long pig," human meat. Fairchild, now blooming as a gourmet, learned that cannibals traditionally baked people with yams and taro roots in pits filled with banana leaves. He was also told that the thumb and palm are the tastiest bits of the body.

Later, on a nearby island, they were introduced to the king of all the Cannibal Islands, a man of higher rank, but that meeting was less dramatic. "Instead of a ceremonial reception such as we had received a few days before, the king was playing cards when we came in and evidently had been drinking heavily," Fairchild wrote. "He merely turned, gave an unintelligible grunt or two, and went on with his game."[12]

*   *   *

The first Lathrop-Fairchild trip, a whirlwind tour that lasted about six months, gave Fairchild a taste of the most exotic ports in the Pacific. He identified a few foods that might grow in America, but he didn't collect any seeds or cuttings because he had no place to send them. The agriculture department had no gardens or laboratories to test tropical plants. Lathrop thought he had a solution, a vague idea that he would establish an introduction garden for fruits and vegetables in Hawaii, so the pair stopped there on the way home. They stayed long enough for Fairchild to see for the first time an avocado, which was then called an alligator pear, but they made no progress toward setting up a garden.

The travelers discussed Lathrop's idea with a few wealthy landowners, but no one was interested in donating property for the experiment. By the summer of 1897, American plant exploration appeared to be nothing more than an ill-defined scheme cooked up by two mismatched vagabonds.

Fairchild was disappointed when he and Lathrop separated after they arrived in San Francisco in August 1897, writing afterwards, "I feared that I had failed him and that I should perhaps never see him again."[13]

In ordinary times, Fairchild would probably have been correct. He would have resumed his mundane government work and, for a few years at least, regaled colleagues with stories of his globetrotting. These times were not ordinary, however. The United States had changed dramatically since David Fairchild had departed on the SS *Fulda* almost four years earlier.

When Republican William McKinley took office as president of the United States in March 1897, he needed to garner political support among farmers, many of whom had voted for his populist Democratic opponent, William Jennings Bryan. McKinley hoped to find a secretary of agriculture whom farmers would respect, one who would allay their fears that the new administration favored bankers over workers. McKinley quickly found the ideal man for the job.

His nominee was James Wilson, sixty-two, a former corn, hog, and cattle farmer who had been born in Scotland and moved to Tama County, Iowa, as a boy. There he was raised, he said later, on "thrift, psalms, and oatmeal."[14]

Before McKinley nominated him as secretary, Wilson had run a farm in Iowa, directed the plant experiment station in Ames, and served three terms in Congress. Midwestern farmers and Washington lawmakers respected him from the first day he took office. McKinley's appointment was so successful that Wilson held his job for sixteen years through three Republican administrations, making him the longest serving cabinet member in U.S. history.

McKinley and Wilson's concern about agriculture wasn't simply political or compassionate; they were also worried about a future food shortage. Statisticians had calculated that if American farmers didn't start producing

FIGURE 5. U.S. Secretary of Agriculture James Wilson in about 1906.
Photo by Frances Benjamin Johnson from the U.S. Library of Congress, item no. 93501342.

more wheat, the nation and perhaps its foreign trading partners faced widespread famine within a generation. Wilson understood that his job was to improve crops quickly. As farmers saw their profits decline from their standard products, they demanded that leaders in Washington tell them about new or little-known alternatives. Wilson needed experts to show him how to help. "He had the firm conviction that what agriculture needed most was more knowledge," Fairchild wrote.[15]

By the time he returned to Washington in the summer of 1897, Fairchild had become an expert, a man who knew more about foreign plants than virtually anyone in town. His travels had transformed his ideas about the food and crops America could grow. "My taste buds were now wide awake," he observed. "I had seen and tasted the food of a dozen tropical peoples."[16]

Despite his new knowledge and enthusiasm, however, Fairchild was unemployed. To save money he moved in with Walter Swingle, who was back in Washington living in a boarding house on California Street between stints of scholarly work in Europe. The two longtime friends, each more confident and a little cocky because they had seen the world beyond Kansas, stayed up late, talking intensely into the night about their futures and the problems facing America. Fairchild learned from Swingle that "the idea of plant introduction as a government activity was germinating in other minds besides Mr. Lathrop's and my own," he wrote.[17]

One evening, as they debated their next career moves, Swingle came up with the breakthrough idea. "Why not organize an office of plant introduction?" he suggested.[18] It was a clear, possibly brilliant solution to the nation's—and Fairchild's—problems. Swingle understood farmers' need for new food from foreign lands. Four years earlier, when he was stationed in Florida trying to protect orange trees from frost and other calamities, Swingle recognized that citrus growers needed different varieties if they were going to stay in business. Speaking to a horticultural society in 1893, he had suggested that growers hire a young man—he mentioned no names—to travel around the world hunting for new oranges to grow in Florida.

"I feel sure a well-trained, intelligent young man could easily be induced to go for his expenses or for a very small additional salary," Swingle told the farmers.[19] None of the financially pressed farmers volunteered to finance the project, so Swingle's proposal went nowhere.

During those sultry nights in downtown Washington, however, the idea seemed feasible if the federal government, not private companies, bankrolled the scheme. Swingle and Fairchild knew that Secretary Wilson was an ingenious administrator who valued scientists' contributions to agriculture. A few weeks after taking office, Wilson had sent an old friend from South Dakota on a quick trip to Russia to gather promising varieties of grains and fruit. Fairchild and Swingle guessed that the new secretary might appreciate their bright idea about how to organize the project.

The two had a clear plan. First of all, they were only interested in new foods and other useful plants, nothing ornamental or impractical. Also, they needed trained botanists to do the hunting so the government wouldn't be inundated with worthless material. Next, they wanted experiment gardens prepared to test the foreign plants. Finally, Swingle and Fairchild proposed, the whole operation could be funded by quietly diverting $20,000 (equal to about $500,000 today) from another line in the agriculture department's budget.

It was an audacious scheme from two junior botanists. But by then Fairchild had grown more confident. "David Fairchild's shyness had disappeared when he had learned to cope with the brilliance and audacity of Barbour Lathrop," observed Marjory Douglas. "After that, he could cope with anyone."[20]

Fairchild and Swingle were apprehensive when they entered their new boss's office at the end of August 1897 even though they had arranged for a senior department employee to go with them to give their idea more credibility. "Secretary Wilson was a tall, gaunt man with a gray beard and deep-set eyes," Fairchild remembered. "He sat listening to us with his eyes half closed and, at intervals, made use of the nearby spittoon. . . . I waited breathlessly for his verdict."[21]

Finally Wilson, who spoke with a Scottish burr, said yes—and he rehired David Fairchild to run the new program, effective immediately. As a nod to members of Congress who counted on the department to give them seeds they could send to their constituents, Wilson named it "the Section of Seed and Plant Introduction." Swingle, who was set to leave Washington to study in Europe, was picked as one of Fairchild's first explorers.

It was a daring decision that had never been tried before. No modern government had employed its own team of full-time plant explorers. In England and France, large private companies had sponsored many foreign plant expeditions to increase their profits by selling rare plants, usually showy ornamentals. These private firms were fiercely competitive and proprietary about their discoveries, but the U.S. government would be eager to share its findings with the public and let farmers make money. Fairchild's plan had exciting implications for President McKinley's constituents.

As a base of operations, Wilson assigned Fairchild a small office four flights up in the attic of the department's overcrowded red brick headquarters. For help Wilson gave him only a secretary named Grace Cramer, a teenager from Kansas still young enough to wear her hair in a long braid down her back. The two were ordered to collect and distribute seeds and plants from around the world. Wilson told Fairchild that, if the shipments crowded the office, he could stash the overflow in an abandoned carriage house out back.

By the end of October, Fairchild was firmly enough established for Wilson to boast about his new program. "A scientist has been appointed in the department to have charge of seed and plant importation," Wilson wrote in his first report as secretary in the 1897 *Yearbook of Agriculture*. "Our country has profited by introducing new seeds and plants, but [until now] much of this work has been done in the dark.... The Old World contains many things that would be valuable to the New World."

For months, Fairchild worked long hours to publicize the new project. He wrote and distributed a detailed bulletin that explained the new office's purpose: "The menu of the average American dinner includes the product of scarcely a dozen plants," he asserted, "and yet the number which could be grown for the table would reach into the hundreds."[22]

Send us your food plants, David Fairchild told the world.

As he hoped, well-meaning foreigners and American travelers began to ship seeds and cuttings to Washington. Fairchild carefully recorded each delivery and distributed samples to gardeners across the country, just as he and Swingle had planned. The new operation began well, but the heavy clerical duties soon overwhelmed Fairchild. His deliverance came during the summer of 1898, after only one year of work, when Barbour Lathrop reentered his life and disrupted it again.

As he done before in Naples and Java, Lathrop suddenly arrived in person as Fairchild was engaged in his valuable but sedentary work. Wasting no time, Lathrop tempted him with the offer of another exciting trip to faraway lands, one that would be longer and more interesting than their six-month cruise through the South Seas. When Fairchild protested that he had just started his new job, Lathrop argued that he was too inexperienced to supervise international plant collectors. If the government's scheme were to succeed, Lathrop insisted, Fairchild couldn't depend on strangers to send the material he wanted. He needed to visit the places himself and make important contacts with botanists, gardeners, and government officials.

Lathrop told Fairchild he was as unqualified to run the office as "a man who had never seen a chicken was fitted to run a chicken coop," Fairchild said later. "His argument was that a world plant collecting service required in it the presence of someone who had seen the whole world."[23]

Again Lathrop won the argument. Despite the commitments Fairchild had made to the U.S. secretary of agriculture and to his colleagues, he decided—after only a few sleepless nights—to quit his new job and follow Lathrop. He wanted to discuss his decision with Swingle, but he couldn't because Swingle was studying in Italy. "I am very anxious to hear what you think of my going off again," Fairchild wrote him on November 7, 1898. "I believe I am not making a mistake for, as you see, it is really now or never."

Wilson did not see the situation that way. He was furious and accused Fairchild of running away from his important responsibilities. "I don't approve of it at all," he said during an unpleasant one-hour meeting.[24] Wilson asked Fairchild to identify someone in the department who could take his place, but Fairchild admitted that he couldn't think of anyone. Wilson was so angry that he refused to pay Fairchild's salary during his travels. Wilson effectively fired Fairchild.

Lathrop rushed to Washington to try to save Fairchild's job by persuading Wilson of the expedition's value to the U.S. government, but the secretary left for Chicago without meeting him. Lathrop followed him there, but Wilson never spoke to him. Despite his fury, Wilson did give Fairchild some help. Although Fairchild was no longer a government employee, Wilson agreed that he could call himself a "special agent" of the department's new section of seed and plant introduction. (That title, which made its bearer sound like a spy, was eventually replaced by "agricultural explorer.") Wilson also supplied ornate letters of introduction to help Fairchild arrange meetings with foreign officials and important scientists. "These papers were wonderful creations, hand-printed on parchment, bedecked with ribbons, and emblazoned with the gold seal of the Department of Agriculture," Fairchild wrote later.[25]

By mid-October 1898, Fairchild, showing no remorse about disrupting his grandiose government scheme, announced blithely in a postcard to Swingle that he was traveling again. Lathrop, Fairchild wrote on October 18, 1898, "came, saw, and conquered."

Tramps Together

The two-year trip Lathrop had promised turned into a five-year odyssey. It was a remarkable adventure of luxury travel experiences, punctuated by meetings with prominent horticulturalists—few were lowly enough to be called gardeners—and casual, dreamlike botanizing sessions on remote islands.

Before they left America, Lathrop showed Fairchild how splendid his fancy life could be, even in America. Their first stop was York Harbor, Maine, then a fashionable summer resort. They were guests of Lathrop's rich sister, Florence Lathrop Page, and her husband, Thomas Nelson Page. The Pages had used a chunk of their Chicago real estate fortune to build a huge cottage on a cliff overhanging the Atlantic coast. They filled it every season with relatives, visitors, and friends who lived nearby. Everybody played golf, tennis, and croquet. The massive dining room table accommodated fourteen guests. The Pages' many friends and neighbors included William Dean Howells, one of the most prominent writers in America. The conversations at the dinner table must have fascinated Fairchild.

His visit to Maine in the summer of 1898 was brief. Because Lathrop was paying the bills, traveling was always conducted on his terms: expensive, comfortable, quick, and not always in a straight line. The zigzagging began immediately after the two men left Maine for California where Fairchild met Luther Burbank, America's first celebrity nurseryman. Burbank had caused great excitement in horticultural circles by inventing startling new varieties of fruits, vegetables, and flowers in these years before scientists understood the science of plant breeding. Burbank's most popular hybrid was the classic Shasta daisy, a flower he created in 1890 that helped earn him the nickname "the plant wizard."

"Burbank is one of the most charming geniuses that it has ever been my lot to meet," Fairchild told Herbert Webber, a colleague in the agriculture department, on December 1, 1898. "He is a marvel, and there can be no mistake about him as possessing the flare which distinguishes a man of genius from an ordinary mortal."

After they left Burbank's spread in Santa Rosa, Fairchild and Lathrop took a quick train ride across the country to New York City, where they spent a few weeks at the swanky Albemarle Hotel on Madison Square, a popular resting spot for Sarah Bernhardt and Lily Langtry. On New Year's Eve 1898, the two travelers boarded a steamer bound for Kingston, Jamaica, the first foreign stop on their around-the-world tour of gardens and food markets.

From the beginning their rapid pace made Fairchild unhappy. "Mr. Lathrop said that ours was not an expedition to exhaust the plant resources of any one island," Fairchild complained. "If I wanted to see the world, I could not spend months in one place."[1]

Lathrop gave him a week to check out the botanical gardens and tropical fruit markets in the Caribbean. That meant, for example, one day in Grenada and a half hour in Haiti. It wasn't long enough to learn anything important, but Fairchild saw enough to inform Swingle on February 19, 1899, of a legend that Haitian merchants sold human flesh in the markets for voodoo ceremonies.

Trinidad, Jamaica, and Barbados received a little more attention. In Kingston, Fairchild first tasted chayote, a mild-tasting squash that he later tried hard to persuade Americans to appreciate. Fairchild collected sixteen varieties of yams and four kinds of sweet potatoes, nutritious stables in the Caribbean diet.

Many officials were gracious and generous, but the two special agents from the U.S. government weren't always treated with the respect that Lathrop expected. In Barbados the pair wanted to meet William Morris, the famous English designer who had been knighted and named agriculture commissioner for the British West Indies. Lathrop contacted him to ask for a meeting, but Morris wasn't welcoming. "He said we could call at three p.m. if we sent a note," Fairchild told Swingle in his February 19 letter. "Lathrop was so angry he would not go near him; insisted, however, I ought to go."

He went, but Morris was chilly. Fairchild, using the confident tone he saved for letters to Swingle, wrote that he had "had one of those uncomfortable times the impolite Englishman knows how to give. I think Morris is nothing short of a conceited English puppy whose head has become actually gaseous with conceit."

From Trinidad Lathrop and Fairchild sailed to Venezuela. They landed during Carnival and again expected notable residents to entertain them, but their arrival was ignored. "The society is very exclusive in Caracas and, though I had letters to the best families, I did not get a single invitation to dinner," Fairchild complained in the long letter to Swingle on February 19, 1899.

Throughout South America, Fairchild hunted for plants the easiest way possible: he bought them in local markets and took cuttings from plants in botanical gardens. At this point in his travels everything was so new and Fairchild's interests were so broad that he randomly collected samples of almost everything that was unfamiliar, potentially interesting, or recommended by local officials.

He shipped large batches to Washington, often without providing information or advice for the people who were supposed to test the plants. By July 1899 the department had received more than two hundred samples of Latin American beans, peppers, squashes, melons, peas, apples, and other fruits and vegetables. Fairchild's most successful discovery during the first part of the expedition was an alfalfa from Lima, Peru, that eventually flourished as a forage plant in Arizona known as the "Hairy Peruvian."

In Chile he bought a bushel of avocado seeds that wound up in California; they produced one of the earliest varieties grown there. Many foods Fairchild collected failed; he admitted that a large percentage of the plants he shipped were lost before they got a chance to grow in America.

During their travels Fairchild and Lathrop faced little physical danger. The closest they came, according to Fairchild's memoirs, was a brief incident when a mule stumbled while carrying him along a mountain path in Argentina. "The enormous height of the Andes, their incredibly steep, rocky slopes and their aridity, made an indelible impression upon me as my mule plodded up the narrow trail," he reported. "When we had almost reached the top, my mule slipped on the ice and suddenly fell to his knees. There was a horrid moment of suspense while he struggled to save himself and me from plunging down a thousand-foot abyss."[2]

While the physical dangers were few, the men were constantly exposed to illness. When they arrived in Panama in February 1899, a few years after yellow fever had forced French engineers to abort construction of the canal there, Panama was considered the most dangerous place in South America. Death was so common that all hospital patients were fitted for coffins when they were admitted for treatment. "Practically none ever emerged alive," Fairchild wrote.[3]

Disease was also widespread in Brazil, the last stop on their South American tour. Fairchild thought Rio was beautiful and apparently agreed with Lathrop that the harbor was one of the three most beautiful sights in the world. (The others were Damascus, which Fairchild never visited, and the volcano in Buitenzorg on Java.)

Many foreigners lived two hours outside Rio de Janeiro in the high-altitude suburb of Petropolis to avoid yellow fever. They knew it was safer than the city, but they didn't understand why. Later it became clear that the elevation was too high for the disease-carrying mosquitoes that appeared at dusk.

When Lathrop and Fairchild arrived in Rio in May 1899, the U.S. ambassador to Brazil was Charles Page Bryan, a relative of Lathrop's from Chicago. "Cousin Charlie" warmly welcomed Lathrop and his scientist companion. One evening Bryan took them to a lavish dinner at the estate of Eduardo Prado, the owner of a large coffee plantation and publisher of a big Brazilian newspaper. Lathrop enjoyed himself tremendously, dining and chatting with stylish, important people who accepted him as their equal. During the meal, Lathrop scrawled on the back of a menu his philosophy of life and presented it to Fairchild as a souvenir. "A moment of great pleasure," Lathrop scribbled, "is worth more than years of dull monotony."[4]

After their ride across the Andes, Lathrop caught malaria. In May 1899 the pair left Rio by steamship so that Lathrop, like many Americans during travel's golden age, could recover while taking the waters in Carlsbad,

Germany. Fairchild was stuck in Germany all summer—except for a brief trip to London at the end of July—and discovered only horse radishes and pickles to send to America.

By autumn, Lathrop was well enough to travel to Italy, but he was grouchy because he believed that U.S. officials didn't appreciate him and his efforts to collect foreign foods. Since Secretary Wilson had refused to meet him, Lathrop couldn't complain to him directly, so he instead groused to Walter Swingle, who had returned to Washington. "I entered into the work with an enthusiasm, which was not in the least chilled by a severe attack of malarial fever brought on by that work," Lathrop wrote on October 2, 1899. "But it has always puzzled me that no one in the department sent me a line of acknowledgement of services rendered, money expended and plants shipped, to say nothing of any appreciation shown for the subsequent work expected of me."

Lathrop had paid all Fairchild's traveling expenses as well as the cost of sending about 250 foreign foods and plants to Washington. He expected the federal government to reimburse him only for the purchase prices of the plants, a sum that by October 1899 totaled $338.50. Despite receiving several vouchers from Lathrop, department officials had ignored his request for payment. Lathrop was, he told Swingle in his letter, "hurt that my efforts for the country and for the department have been so absolutely unappreciated or misjudged by the officials of the department."

By the time the pair arrived in Italy in November, Lathrop was so angry that for several weeks he refused to send any more plants. During this period, however, the agency received anonymously many Italian plants that Fairchild probably mailed behind Lathrop's back. These secret shipments included broccoli, then virtually unknown in America.

In Venice Fairchild also discovered zucchini—identified as "vegetable marrow"—for sale in a market. He had first heard about the squash during his quick trip to California in 1898, but the food was so unknown in America that Fairchild felt obliged to include advice on how Venetians cooked it: "fried in oil and tomatoes; fried with eggs, etc., much as eggplants are treated. Said to be of very delicate flavor."[5]

Later that month, the two men sailed from Naples to Cairo, a city that opened Fairchild's eyes to more new foods and farming methods. Before he arrived in Egypt he said he knew the word *sesame* only as Ali Baba's famous password; afterward he understood it to be a source of valuable cooking oil. He also collected chickpeas, okra, strawberry spinach, and more hot peppers. In Cairo they stayed, like most wealthy foreigners, at Shepheard's Hotel, one of the smartest hotels in the world, while Fairchild toured the Khedival Agricultural Society's grounds and learned about Egypt's important cotton industry. "As Mr. Lathrop and I were drinking our coffee on the verandah of Shepheard's Hotel and watching caravans of camels and donkeys passing on their way to and from the desert, I described the cotton situation and told him of the samples I was sending to Washington," he wrote.[6]

After his pique finally cooled, Lathrop encouraged Fairchild to buy as much cotton as possible. He shipped six bushels of seeds of three varieties, material that eventually boosted the lucrative cotton industries in Arizona and California.

In January 1900, perhaps in an attempt to appease the grumbling Lathrop, the agriculture department published a special list of hundreds of plants Lathrop and Fairchild had found in Austria, Italy, and Egypt. In the inventory's introduction, Orator F. Cook, a government botanist who was filling in for Fairchild in Washington, publicly acknowledged the shipments had been "received through the generosity of the Honorable Barbour Lathrop of Chicago."

His mood was improved by Cook's comments, and Lathrop took Fairchild back to Java to start the most romantic part of their expedition. It was a twenty-day tour of eighteen Pacific islands using Alfred Russel Wallace's best-selling book, *The Malay Archipelago*, as their guide.

As their ship cruised slowly from island to island in January 1900, Fairchild rushed ashore to botanize every chance he had. Romance got the better of him, however, as he collected and sent to Washington a batch of plants that were not exactly useful: the forty-nine items he collected included twelve different red peppers, two mysterious plants from a mangrove swamp, an unidentified fruit that wasn't edible, and, best of all, samples of a sap that natives used to poison arrow tips. Secretary Wilson must have been delighted with the shipment.

Banda, a small cluster of islands in the Moluccas, a part of the Malay

Archipelago, captivated Fairchild. "I have been to many out-of-the-way places, but I think Banda seemed more out of the world than any other spot I have ever visited," he wrote later. When Fairchild arrived, Banda was an important source of nutmeg, an especially handsome plant. "There are few fruit trees more beautiful than nutmeg trees with their glossy leaves and pear-shaped, straw-colored fruits," he recalled. "As the fruits ripen, they crack open and show the brilliant crimson mace which covers the seed or nutmeg with a thin, waxy covering. The vivid color of the fruit and the deep green foliage make the trees among the most dramatic and colorful of the tropical plant world."[7]

Fairchild, who rarely passed up an opportunity to stroll alone among trees, spent hours wandering through nutmeg groves. He was so enraptured by the wonders he saw that one tiny island's sights overwhelmed his common sense. He landed at Boela, a village on the island of Ceram that the Standard Oil Company was developing into an oil center. It was a beautiful spot. "A beach of pure white sand shaded by graceful coconut palms surrounded by the calm waters of the lovely lagoon," he reported. "At the edge of the water, picturesque thatched cottages of bamboo stood on piles to raise them above the waves of any storm which might sweep across the island. Wandering among the houses were naked children and women in batik dresses of Javanese design. The atmosphere of the place was completely tranquil and carefree."[8]

As Fairchild landed on the beach, he was greeted by a series of fantastic sights. Hundreds of six-inch-long fish were jumping in the surf. Tiny hermit crabs scuttled along the sand. A small, hairless dog with a yellow topknot barked at him. Finally, two chickens with bright blue plumage strutted across his path. "This combination of novelties quite paralyzed me with astonishment," he recalled. "I was convinced that I was seeing sights which no other naturalist had ever beheld."[9]

Fairchild had let his romantic notions about exotic travel make a fool out of him. He learned later that the fish, the crabs, and the dog were creatures well known to many westerners and that an American oil driller with a sense of the absurd had dipped the chickens into blue dye.

Not all the stops they made on the circuit were romantic, however. In Fakfak, New Guinea, they met unfriendly natives, an experience that unnerved Fairchild. "While I enjoyed the novelty of looking into the faces of primitive men, the fact that I could not speak a word of their language and knew that they were distinctly hostile gave me a decidedly unpleasant feeling," he wrote later.[10]

Despite his discomfort, he was curious to know what the natives of New Guinea discussed. He learned that their favorite topics were familiar ones. "I found that the Papuans were always talking about food and one another and about clothing, though they wore next to nothing," he wrote.[11]

After leaving Indonesia, Fairchild and Lathrop sailed to the Philippines, which had become a U.S. possession after the end of the Spanish-American War (and, more important to Fairchild, a potential spot to experiment with tropical food plants). In March 1900, they went to Hong Kong where Lathrop found luxurious lodging while Fairchild enjoyed a short solo side trip into China. The excursion transformed his ideas about foreign plant exploring.

Fairchild took a seventy-five-mile boat ride up the Pearl River to Canton, one of only five Chinese ports that officials allowed foreigners to visit. There he glimpsed a vast floating city that was home to half a million people. Seaman A. Knapp, a colleague of Fairchild's who had visited Canton the year before, described the people he saw. "They are born, marry and die in their boats, which cost an average twenty-five dollars and last some fifteen years," he wrote in his diary.[12]

Canton's land was crowded, too. Fairchild was overwhelmed by the teeming city's congestion. "I shall never forget the canals, sampans, truck gardens, pagodas, tiny orchards and thousands upon thousands of hanging shop signs decorated with Chinese characters," he remembered.[13]

Lathrop arranged for an English-speaking guide to escort Fairchild so he could catch a few glimpses of farms outside Canton. For Washington Fairchild collected four rice varieties, an unnamed fruit he found in the neighborhood called the City of the Dead outside the city's north gate, and an example of the myrobalan, a sacred fruit used in Ayurvedic medicine.

These plants probably didn't impress Secretary Wilson, but the quick trip had an enormous influence on Fairchild. He vowed to find a way to explore China's vast lands and penetrate its mysteries. "I returned to Hong Kong with the feeling that I had been living in a dream," he wrote later.

"Surely only a nightmare could fill my brain with such fantastic people, practices and customs. Everything which I had experienced during my days in Canton seemed utterly unbelievable."[14]

While he was touring Canton mesmerized by the botanical potential of China, Fairchild wrote to Augustine Henry, a Scotsman working as a civil servant for the Chinese government. In his spare time during the almost twenty years he spent in China, Henry had taught himself enough about Chinese plants to become a celebrated amateur botanist. He had discovered hundreds of new plants during his collecting trips, primarily around Ichang, an important port on the Yangtze River that would later play a significant role in the story of America's plant explorers.

Fairchild asked Henry's advice about the best way to obtain Chinese plants for America. Could he depend on western missionaries to collect material and mail it to Washington? Henry discouraged him. He said plants in China were as valuable as gold in the Yukon, but finding treasures would be difficult. "American enterprise might accomplish something," Henry wrote to Fairchild on March 16, 1900. "But I wouldn't waste money on postage; send a man."

Fairchild took Augustine Henry's advice seriously.

* * *

Lathrop and Fairchild continued moving around the world for another four months. They sailed from Hong Kong to Ceylon, a significant crossroads for steamship traffic, where Fairchild intended to hunt for plants in the mountains above Kandy. The day after they arrived, however, it was Fairchild's turn to get sick. He contracted a severe case of typhoid fever. His condition terrified the manager of the Kandy hotel, who insisted that the patient leave before he infected other guests.

"Mr. Lathrop had seen the little hospital to which I would have to go and was far from satisfied with it," Fairchild wrote later. "When the manager persisted, Mr. Lathrop produced his revolver and laid it on his bureau, announcing that if they attempted to move me, he would shoot."[15]

The threat worked. Fairchild stayed in bed at the hotel under Lathrop's guard until he was strong enough to move to a ship; he recovered on the long trip to England. In May 1900, Fairchild visited Scandinavia to collect examples of tough-weather fruits and fodder plants that might survive the extreme climate of America's northern plains.

When the pair arrived in London in June to prepare for the voyage home, Fairchild finally acknowledged that he had created a mess in Washington by leaving with Lathrop without finding a replacement to do his job. "Fairchild's going away . . . rather left things in bad condition," Beverly Galloway, Fairchild's original mentor, had confided in Swingle on February 20, 1899. "It rather gave the secretary a bad impression of the confidence he could place in scientific men." Nonetheless, Fairchild refused to admit that the mess was completely his fault. "Regarding department affairs," he wrote to Swingle on June 17, 1900, "this is no time to say anything further than that they seem to be in a nasty muddle and there has been probably blame all around."

He deflected criticism further by describing Lathrop not as a passive real estate investor with a savvy brother but as a shrewd capitalist. "Think what the effect of this petty squabbling among we botanists must have upon a businessman like Mr. Lathrop!" he wrote. "You fellows have helped to kill the goose which had made preparations to lay a solid golden egg."

Confident of his own value to the U.S. government, Fairchild rushed to Washington after he arrived in America at the end of August. He learned quickly that—as Galloway had predicted—Secretary Wilson had not forgiven him for abandoning his responsibilities two years earlier. Lathrop tried to help by writing Wilson a letter filled with overblown praise of Fairchild. "No better man could be found, I doubt if any so good, for the head of division of your department," he wrote on November 14, 1900.

The approach didn't work. Again Wilson wouldn't agree to meet Lathrop, and he refused to rehire his protégé. Fairchild's future—and the future of government-financed plant exploring—looked dim. Within days, however, Fairchild's situation improved, thanks again to good luck and an older man's patronage.

On September 11, 1900, William Saunders, a long-serving federal official who supervised the agriculture department's grounds and gardens, died suddenly; his death forced Secretary Wilson to make major changes in the agency. Within days Wilson promoted Galloway to run a new and better-funded division in the department called the Bureau of Plant Industry.

Galloway, suddenly with enough authority to finesse his way around Wilson's irritation, found a temporary assignment for Fairchild: go to Germany to collect information about hops for American beer makers, a plan that kept Fairchild employed for one year. He completed the assignment in Germany and, as his mandate from Washington broadened, roamed through the Mediterranean region looking for new dates and other plants. He was in Alexandria on March 18, 1901, when he got a telegram announcing that his father was dead. He was upset, but there was nothing he could do. "I had no alternative but to push on," he wrote later. "Life is nothing but pushing on, after all."[16]

After he returned to America in August 1901, he found himself—again—without a job.

*   *   *

While Fairchild was touring Europe and North Africa for Galloway, Lathrop had continued his circumnavigations alone. He spent the winter of 1901 socializing with friends in Hawaii, including hosting a fancy dress ball for fifty guests in Honolulu. Despite constantly surrounding himself with people, Lathrop was lonely. The summer of 1901 he wrote Fairchild a long letter arguing that taking another trip together would be better for his career than settling down at a desk in Washington.

"Enduring scientific fame, whether in great or in smaller degree, can only come from thoroughly acquired and properly applied knowledge," he wrote on September 10, 1901. Fairchild hesitated because there had been friction between the two men. Lathrop's rudeness and sarcasm were difficult to endure, Fairchild admitted later. His "unbridled tongue at times drove me as a young man nearly distracted," he wrote.[17]

Not wanting to drive Fairchild away, Lathrop apologized for being a difficult traveling companion who had reproached Fairchild for minor failings. "I want to disabuse your mind of any gloomy thought of possible scoldings in the future," he wrote on October 11, 1901. "Really, there will be many fewer of them on our coming trip than there once was."

Lathrop's pleading sounded pathetic. He was no longer an officious boss, ordering Fairchild around and lecturing him on how to dress and behave in public. Lathrop discovered he needed Fairchild as much as Fairchild needed him.

6

From Far East to Mideast

Lathrop's arguments worked, mostly because Fairchild had no other offers. Lathrop was so pleased that he rewarded his companion with especially generous gifts: a new steamer trunk, a silk hat, and a $250 check—worth almost $7,000 today—"to guard against shortage of funds."[1]

By then, November 1901, each man had changed. Lathrop, fifty-four, was growing old and sickly. He had developed an ulcer, a condition that intensified his natural grouchiness. In contrast, Fairchild, thirty-two, had gained confidence as he learned the important skills of the job—to identify which strange fruits and vegetables had the biggest potential to be useful and how to pack material to survive the trip to Washington. The two travelers had almost become equals.

Fairchild left San Francisco on his second around-the-world expedition on November 16, 1901. Walter Swingle and Swingle's new wife, Lucie, saw him off on the SS *Hong Kong Maru*, a recently launched 6,000-ton ship. The first stop was Honolulu where Jared Smith, a former colleague in Washington, was running an experiment station. As they discussed their work, Smith warned Fairchild that Secretary Wilson still had no interest in tropical plants.

Although Smith was based in Hawaii, Wilson had ordered him to concentrate on familiar, practical foods. Smith should "confine his experiments to such economic plants as might be of interest to the middle-class people of Hawaii," Fairchild said Wilson told Smith. "He was not to study plant introductions even though they might later contribute to the prosperity of the islands. Corn and tobacco were on the program, but the mangosteen, litchi, and mango and the like were not."[2]

Fairchild ignored Smith's advice. Still fascinated by China, he rushed back to Canton as fast as an early twentieth-century traveler could. Leaving Lathrop alone in Hong Kong to nurse a bad toothache, Fairchild again took a seven-hour steamship ride up the river into the teeming, floating city. Among the first foods he sent to Washington were two types of litchis. Fairchild was thrilled to return to China at a key moment in the nation's history.

The Boxer Rebellion, a Chinese uprising against European imperialism, had occurred during the twenty-one months Fairchild had been away. The revolt was put down by troops from eight foreign nations, each eager to exploit China's riches. Under the peace treaty signed September 7, 1901, the foreigners extracted huge sums in retribution. Money to pay the penalties was raised in an unusual tax that increased Chinese suspicion and resentment of people they called "foreign devils." Fairchild told Wilson the government raised funds for the war indemnity by collecting five cents a year on every rafter of every house in China.

Despite this tax burden, the Chinese treated Fairchild well and he had time to introduce himself to John M. Swan, a doctor at a missionary hospital in Canton who helped him collect dozens of peaches, plums, persimmons, and other fruits. Swan also told him how to find the seeds that produce tung oil, the glossy material used to waterproof the exterior of Chinese junks.

With Lathrop stuck in a hotel room, Fairchild was able to visit rural areas outside Canton and wander among the small vegetable plots there. "These truck gardens of a city of 2,000,000 people did not contain a single vegetable with which we are familiar in America," Fairchild reported. "The people, apparently well nourished, live on an entirely different diet from that which we consider necessary for health and happiness."[3]

He saw water chestnuts growing in ponds and swayback pigs with enormous bellies being raised for bacon. He watched Chinese farmers control pests the old-fashioned way: they picked off each insect on every plant by hand. Fairchild's interest and curiosity were insatiable, but when Lathrop recovered, he insisted that his protégé return to Hong Kong.

Fairchild was unhappy. "Those days in Canton were annoyingly short," he complained.[4]

*   *   *

After Hong Kong, the pair traveled to Singapore, Ceylon, and Bombay, where Fairchild separated from Lathrop and began his next ambitious project, exploring Arabia for date palms.

When Fairchild arrived in Karachi from Bombay in January 1902, officials told him that he needed to be vaccinated against smallpox before he could travel through the Persian Gulf region. Fairchild went to the home of a local doctor named Singh to receive the injection. "He could not vaccinate me, he said, until the following day," Fairchild remembered. "He did not have his tools."

When he returned, Fairchild saw the doctor's implements waiting for him. "A calf was standing in the yard," he wrote. "Part of Singh's equipment was a hinged-top table to which he strapped the calf securely and tipped it on its side." Gradually, as Fairchild watched the doctor prepare, he understood what Singh was doing. "I realized that he intended to vaccinate me directly from one of the scabs on the belly of the calf," he wrote, "and I began to feel very unhappy."[5]

Fairchild, always polite and open-minded about foreign ways, allowed the doctor to vaccinate him, but he foolishly disinfected the wound when he returned to his hotel room. Dr. Singh's primitive method had successfully inoculated Fairchild; distrusting the doctor left him unprotected against smallpox. Yet Fairchild was lucky, as usual, and didn't get sick.

On this voyage through the Persian Gulf, Fairchild met a dashing British soldier, identified only as Lieutenant W. H. Maxwell, who was returning from Peking, where he had helped put down the Boxer Rebellion. Because he was an officer in a light infantry division, Maxwell enjoyed many perquisites of rank and traveled with his own polo ponies. As they chatted, Fairchild asked him about foods he had eaten during his deployment. Maxwell described his favorite discovery, a delicious nectarine he had eaten in Quetta, a city now in Pakistan. Maxwell later shipped seeds of the fruit, which is actually a smooth-skinned peach, that thrived in California.

Fairchild was especially proud of the introduction's remote origin, the land of the Afghans.

From Karachi Fairchild took the SS *Pachumba* on a thirteen-day trip to Basra, now a province in Iraq, making several stops along the way. One was Muscat, a city now in Oman that was an important date palm center. The weather wasn't bad when he arrived in February, but the ship's captain told him that in July, when temperatures reached as high as 115 degrees, he and his officers kept cool by lowering the gangway into the sea and sitting in the water all night. Fairchild also stopped in Jask, now a city in Iran, a desolate spot. "The coast of the Persian gulf is so barren that one can travel for many days along it without finding so much as a single tree or shrub," he wrote later.[6]

Fairchild did notice a remarkable sight in the treeless plain at the junction of the Tigris and Euphrates Rivers: date boats. They were built entirely from the midribs of palm leaves, which measured ten feet long and were secured with wooden pegs and strong twine.

Fairchild's own ship was filled with an exotic assortment of people: Arabs, Persians, and Hindus, including one Brahmin who did not eat anything on the voyage because, Fairchild explained, "there was no suitable place to prepare it, no spot where during its preparation the shadow of some infidel might not fall upon the food."

Fairchild, who slept on deck, woke early every morning to the prayers of a Shiite pilgrim on his way to Karbala, a holy city southwest of Baghdad. "It seemed to me that he repeated the name 'Hosain' at least a hundred times," he wrote later. "What effects these prayers to Allah had upon his private life I do not know, but whenever I hear the word 'devout,' the morning prayers of this old Shiite pilgrim come back to me."[7]

On the next leg of his two-month voyage, Fairchild traveled from Basra to Baghdad on an unlikely vessel, an old Mississippi River boat that had been shuttling up and down the Tigris River for years. "Nobody seemed to know how such a craft had reached Mesopotamia, but there she was," Fairchild recalled.[8]

In 1902 Baghdad, the dream place Fairchild learned about in front of his fireplace in Kansas, was also full of surprises. The streets were narrow and dusty, and the houses were covered with latticed windows that hid harems from public view.

Fairchild arrived about seventy years after plague had swept through the city, causing so many deaths that residents still talked about the devastation. One told Fairchild that the disease had killed 40,000 people and so terrified survivors that dead women lay in the street with gold bangles untouched on their arms.

While he was in Baghdad, Fairchild, who frequently compared his own adventures to the books and spectacles he had known growing up, witnessed a dance by authentic whirling dervishes, not the circus performers he had watched as a child.

By the time Fairchild finished this two-month detour to the Persian Gulf he had collected 224 date palm offshoots or suckers, each weighing about thirty pounds. He believed they could grow under the similar growing conditions of southern Arizona and California, then generally a region of useless desert. Near Baghdad he met Sheik Abdul Kader Kederry, the largest date grower in Baghdad, who was especially gracious in light of an encounter he had just had with an adversary. "A Turkish soldier had shot him through the arm the day before," Fairchild wrote later, "but he received me cordially, promised to assist me in every way and offered me anything I wanted from his date plantations."[9]

After he arranged to send almost four tons of trees to Washington, Fairchild retraced his route and joined Lathrop in Japan in the summer of 1902. They lived comfortably at the Imperial Hotel in Tokyo where Lathrop relaxed and Fairchild searched for plants. He bought fruits and vegetables at public markets and discovered zoysia, a plant that eventually became a popular ground cover in America.

At Lathrop's insistence he also bought bamboo plants, a purchase that triggered Fairchild's long love affair with this huge grass. Lathrop believed that bamboo was one of the most useful plants in the world and that Americans would eventually recognize its value. "It may take a long time before Americans learn how to use it," Lathrop said, "but they'll never learn if we do not introduce the plant."[10]

Also in Japan Fairchild became as interested in flowering cherry trees as Lathrop was in bamboo. In 1901 he had tried to arrive in the country when the trees were in bloom, but he was a few days late. Instead of seeing the real thing, Fairchild settled for studying beautiful watercolors of the blossoms with an artist named Takagi in Tokyo. "We sat together on the spotless matting in his simple little house and he explained the drawings to me one by one," Fairchild wrote. "I have rarely been so thrilled, for I had had no idea of the wealth of beauty, form and color of the flowering cherries."[11]

Guided by Takagi's paintings, Fairchild picked thirty varieties and shipped them to Washington, where the trees were virtually unknown. Officials sent them to California, but workers there didn't know how to handle them and all the trees apparently died. Despite this failure, Japanese flowering cherry trees remained one of Fairchild's passions.

By November, Fairchild and Lathrop were back in Naples preparing to tour Africa when, at the end of the month, they received good news: James Wilson had finally given Lathrop the public acknowledgment he craved. In his written report in the 1902 *Yearbook of Agriculture*, Wilson boasted about the department's successful introduction of foreign seeds and plants. "This work has been greatly aided by the generosity of Hon. Barbour Lathrop who, at his own expense, has carried on extensive agricultural explorations during the year, assisted by Mr. David G. Fairchild, an agricultural explorer of this department," Wilson wrote. He mentioned bamboos, dates, and clovers as particularly valuable finds.

Buoyed by praise, Lathrop and Fairchild cruised around Africa for six months, stopping at port cities along the way and enjoying lavish meals. Lathrop considered himself a genuine gourmet—the only foods he would not eat were ancient Chinese eggs, chicken embryos, and durian, the smelly fruit—and he enjoyed trying to identify various ingredients in complicated dishes.

Nonetheless, the trip grew tedious as it followed an uninterrupted routine of two days at sea for every one on land. Fairchild said only one spot was memorable. "As we neared the island of Zanzibar, the air was filled with fragrance, for the clove industry, transplanted from the Spice Islands of the East Indies, had already begun to be profitable there," he wrote later.[12]

He found little to collect during the forty-nine-day voyage around southern Africa, although he was happy to spend a week ashore at Durban, then the capital of Natal, one of four states in the Union of South Africa. Despite a lifetime of travel, Fairchild always considered himself a landlubber who became seasick and accident-prone whenever he got on a boat.

"I found it a great relief to look at a horizon not forever shifting its levels, and to sleep at night without being continually tossed from side to side in a narrow bunk," he remembered.[13]

Fairchild discovered a delicious miniature pineapple from Natal and many fodder plants unknown in America. In Cape Town he noticed a shrub called Spekboom. Peter Macowan, an English botanist there, said that it was a favorite food of elephants. Fairchild, who had perhaps been away from Washington too long, sent cuttings to the department with an enthusiastic note about how much elephants loved to eat it. Beverly Galloway, his friendly mentor in the department, responded, "It's all very well to introduce an elephant fodder tree, but I'd like to know where the elephants are to come from."[14]

* * *

During his travels with Lathrop, Fairchild constantly hunted for varieties of one particular food, the mango. It was his second favorite fruit after the mangosteen, which, despite its name, is not related.

Mango trees were rare in the United States when Fairchild was growing up, but by 1903 a few Americans had traveled enough to taste the best varieties and were determined to grow them at home. Henry Perrine, who discovered the fruit when he was U.S. consul in Mexico, was probably the first to attempt growing mangos in Florida; his efforts failed miserably.

America's second known mango-lover had more success. He was Elbridge Gale, a horticulture professor from Fairchild's alma mater, Kansas State College. After Gale retired in November 1884, he and his wife moved to a homestead on the west side of Florida's Lake Worth. The Gales called their new property "Mangonia" and planted all the varieties they could find. Five years after the Gales settled in Florida, a botanist from Pune, India, sent eleven mango trees that were grown from grafts and labeled as Mulgoba to the U.S. Department of Agriculture. The agency gave at least three to Gale, who dared to plant them even though the East Indian mango had a terrible reputation in America. "This was at that time the laughing stock of all who ate it, compared as it was in flavor and texture to a ball of tow soaked in turpentine and molasses," Fairchild wrote later.[15]

Gale's three trees did well until the winter of 1894–95, when they were severely damaged in what became known as the Great Freeze. Only one mango tree survived, but by 1898 it bore enough fruit to attract local mango lovers' attention. It was the first grafted mango cultivated in America and, unlike mangos then grown from seeds, its fruit was luscious, without the famous turpentine taste. It was soft and juicy without much fiber.

Fairchild ate one of Gale's mangos during his first visit to Florida in 1898. It was so delicious that he vowed on the spot that he would persuade Americans that mangos were worth growing and eating. "Many people think they know what a mango tastes like because they have eaten some fruit by that name sold in one of the fruit stores of our cities," Fairchild commented. "The fruits that are offered now as mangos are unworthy of the name, for they are from worthless seedling trees and are little more than juicy balls of fibers saturated with turpentine, while the oriental mango is a fruit fit to set before a king."[16]

Later in 1898, as Fairchild was organizing the office of seed and plant introduction, Elbridge Gale proudly sent him one perfect Indian mango. Eager to prove how delicious the exotic fruit was, Fairchild presented the gift to Secretary James Wilson. "I can see him now as he lifted it from my hand and raised it to his nose with the remark, 'It's got a curious smell, hasn't it?'"[17] He learned later that his gift was not well received. After one whiff, Wilson gave Fairchild's mango to a friend. His friend took it home to his family. They gave it to their gardener. He gave it to his chickens, but nobody would eat it. "They all declined to like it, my precious and only mango," Fairchild complained.[18]

Fortunately, not everyone felt that way. A few months after Fairchild's mango misadventure, Captain John J. Haden, a U.S. Army officer and Civil War veteran, retired with his wife Florence to Coconut Grove, Florida. Soon after the couple arrived, Captain Haden sailed north to West Palm Beach to buy a half dozen of Gale's famous mangos. He planted the seeds on his property and, although Haden died five years later, his trees thrived and produced fruit in 1910. Florence Haden sent one of the first fruits to Fairchild, and this time officials in the agriculture department welcomed the gift.

Fairchild took the beautiful mango—the skin was tinged with scarlet over a background of gold and green—to William A. Taylor, a colleague and fellow fruit lover. Taylor was impressed. "Fairchild," he said, "I believe that is the most beautiful fruit of any kind I have ever seen."[19]

Florence Haden named this gorgeous variety after her late husband and marketed the fruit in Florida, even publishing her own mango chutney recipe in a local charity cookbook. In a genealogical chain similar to a thoroughbred racehorse pedigree, the Haden mango begot a seedling called the Tommy Atkins, which one hundred years later had become the most popular commercial variety in the world.

It was Elbridge Gale's determination and defiance of conventional, wrong-headed wisdom that inspired Fairchild to search for mangos all over the world. During the four years he spent traveling alone and with Lathrop, Fairchild sent twenty-four varieties from six countries, each supposedly tastier or hardier than the other.

When he was in Saigon in April 1902, Fairchild was thrilled to learn that the Cambodiana mango, an excellent variety grown from seed, was in season. He immediately rushed to the closest market and bought every one he could find, about 100 fat fruits. Next he purchased scrubbing brushes, packing material, and lots of charcoal to protect the seeds in transit. Back at his hotel, he asked the head porter to locate half a dozen boys to help him clean the fruit. The kids ate the mangos and scrubbed the seeds. Fairchild packed them in charcoal and shipped them to America where they were eventually used for breeding purposes.

In India in January 1902 Fairchild bought cuttings of what is probably the world's most desirable mango, the sweet yellow variety called the Alphonso, as well as several large baskets of other types that looked promising. Again, he didn't have much time to clean the fruit and extract the seeds for shipment to America, so he and Lathrop used the same labor technique.

On the day the two men were leaving Bombay, Lathrop rounded up children playing on the dock and ordered them to eat the mangos immediately and spit out the seeds. He egged them on by tossing pennies their way as Fairchild dashed around collecting the sticky, slippery seeds. All the while the P&O steamship hooted impatiently to hurry the laggard passengers. The men and the mango seeds made it to Washington. In time, farmers learned to grow varieties that Americans love as much as David Fairchild did.

*   *   *

The Fairchild-Lathrop odyssey finally ended in the summer of 1903. As usual, Lathrop returned to Maine to visit his sister and her family, while Fairchild went to Washington to try to get his old job back. It wasn't easy despite—or perhaps because of—pressure from his mentor.

Again Lathrop had written to Secretary Wilson, somewhat imperiously asking him to find work for Fairchild. This time he got an answer, but it wasn't encouraging. "I note what you have to say in regard to having finished your explorations and your desire that Mr. Fairchild be given an opportunity in the department for future work along the lines of plant introduction," Wilson wrote on June 12, 1903. "I shall be very glad to give this matter careful consideration."

Before he decided what to do with David Fairchild—the eager young botanist who had sent him elephant fodder grass and poison arrow sap—the secretary emphasized to Lathrop that he was interested only in plant exploring that had clear, practical goals.

"I am more and more convinced that the miscellaneous and general introduction of seeds and plants can have very little material effect on the agricultural welfare of the United States," Wilson wrote in his June 12 letter. "Our future success in this country will depend largely on concentrating upon a few things and giving full consideration to these things from all standpoints."

Wilson made it clear that from then on Fairchild was going to have to seek romance in Washington, not in exploring for exotic foods in faraway lands.

The Ends of the Earth

James Wilson was pleased by the work of his personal explorer, one of the first appointments he made as U.S. secretary of agriculture. To Wilson, Plant Explorer Number One was Niels Ebbesen Hansen, a horticulture professor at South Dakota State Agricultural College. Hansen was always Wilson's favorite. "I have twelve thousand men under me, but none who knows how to work like Hansen," Wilson said. "There is only one Hansen."[1]

Hansen, who emigrated from Denmark when he was seven years old, was a young plant breeder who worked in the northern plains, the region that Wilson was trying hardest to help. Hansen had done some traveling before Wilson hired him in spring 1897, having visited Russia and seven other countries for four months in 1894 while he was a student at Iowa State College and Wilson ran the plant experiment station there.

Hansen also had another, more important qualification for the job. Unlike many other horticulturalists at the time, he was a plant breeder who understood that it was botanically impossible to acclimate plants to tolerate severe conditions; only cross breeding with proven hardy varieties could produce tough plants. Because Hansen possessed this scientific sophistication, Wilson trusted him to know what to look for in the field.

Hansen was thirty-one in 1897 when Wilson convinced him that the future of American agriculture depended on his returning to Russia to find material that could be introduced in the Dakotas, then a dry, unproductive region where few crops grew. The mission was haphazard and dangerous. Wilson paid him $3,000, a generous salary equal to about $78,000 in current dollars.

FIGURE 6. Niels Hansen dressed for work in Siberia. Permission of South Dakota State Agricultural Heritage Museum Photographic Collection.

At first Hansen was enthusiastic about the assignment and ordered special clothes and equipment for the job. "His exploring outfit consisted of a rubber billy [club], a dagger on his right side, a revolver in his belt, field glasses, and magnifying lenses," his biographer wrote.[2] During one trip to Omsk, Siberia, he posed proudly wearing this garb for his official portrait.

For seven months in 1897 and 1898, Hansen traveled through Russia, Siberia, and China, hunting for commercially promising plants that might survive the extreme growing conditions in South Dakota and nearby states. At first, he traveled like a European diplomat, wearing a tall silk hat and formally presenting his credentials to representatives of the Russian government. When he traveled outside cities in bad weather, however, he resorted to less elegant gear. During a trip to Siberia at the end of August 1897, the air was so dry that Hansen had to wear a wet sponge over his mouth to avoid inhaling dust.

Shortly after Hansen arrived in Uzbek province in Turkistan in November 1897, a field of alfalfa with small blue flowers attracted his attention. He believed the plant would survive in South Dakota, where temperatures range from 50 degrees below zero to 114 degrees above, to provide year-round feed for livestock, as well as produce nitrogen to enrich the soil. Before he could recommend the plant to Secretary Wilson, however, he needed to figure out how far north the blue alfalfa grew.

Jolting through the mountains in a low sled on a route taken by Alexander the Great, Hansen followed the plant's trail to see where it stopped growing. He searched fields, questioned soldiers and natives, and examined bags of horse feed for sale in public markets, collecting alfalfa seeds by hand along the way.

On Christmas 1897 he reached Kopal in southwestern Siberia, a town on the same latitude as South Dakota, where the blue Turkistan alfalfa was still growing. Confident it could thrive on the Northern Plains, he sent thousands of seeds to Washington. (Years later he returned and discovered a hardier type, an alfalfa with tiny yellow flowers, and brought that one to America, too. As a lasting tribute to Hansen's work, South Dakota State University selected blue and yellow as its school colors.)

Hansen covered 2,000 miles on his search, 1,300 by troika and 700 by sleigh. On most of the expedition, language was a problem. In Siberia he needed three interpreters: one to translate from Chinese to Tartar, a second from Tartar to Russian, and a third from Russian to German. Hansen spoke the last language fluently.

The return trip from Kopal at the end of 1897 was hazardous, almost fatal. Hansen was prepared for cold weather—he wore a one-piece double reindeer suit with fur on the inside and outside—but he almost froze to death one night after his carriage broke down in a blizzard with temperatures of 50 degrees below zero and he had to sleep outdoors until the weather cleared. "Furious storm, wind and snow today," he recorded in his diary on December 29, 1897. Conditions were so treacherous that Hansen wrote a long letter to Secretary Wilson telling him how to handle his personal belongings if he died. Hansen had apparently caught strep throat during the storm, but he had to travel all night through another blizzard to find a doctor. "Was nearly gone by the time I reached the hospital," he wrote in his diary. "Doctor called twice." While Hansen was recovering at a military hospital in Sergiopol, he came close to dying a second time when dangerous fumes from a low fire in his room almost asphyxiated him. "In evening nearly suffocated from stove leaking gas," Hansen wrote. "Called the doctor again."[3]

He spent his thirty-second birthday in a hospital during yet another blizzard. "Would that I could celebrate it at home," he noted sadly. By the time Hansen recovered and was safe in a decent hotel room in Semipalatinsk, Siberia, he confided in a letter to his future wife that he had faced great difficulties. "It has been a much harder trip than I anticipated," he admitted. "I found many interesting plants, but would not have undertaken the trip if I had known how hard it would be."[4]

* * *

Hansen was halfway through this expedition when James Wilson appointed David Fairchild to coordinate foreign plant introduction. As Fairchild was setting up the new division, huge shipments of Hansen's material began arriving at his tiny office. Hansen sent about 750 types of seeds and plants, including winter melons from Turkistan that plant breeders in California and Utah eventually transformed into the Persian and the Honeydew. At first the parcels trickled in from Russia; soon, however, hundreds of packages arrived in a deluge. One day in February 1898, twelve tons of seeds of a fodder plant called smooth brome grass from the Volga River district turned up.

Fairchild struggled to keep the shipments straight and check for dangerous insects or diseases that might have accompanied the material. The department had organized a system of public and private experiment gardens to test the material, so Fairchild arranged the seeds into 5,000 small packages and shipped them around the country. The enormous workload made him miserable. Fairchild, who hated clerical tasks, soon decided that he would rather be exploring himself. Again he was unhappy. "Hansen felt that he had been sent out to collect, and he collected everything and collected it in quantity," Fairchild recalled. Later in an unpublished essay his criticism was harsher: "Hansen's collections took on the character of a nightmare."[5]

Nonetheless, Hansen had Secretary Wilson's support, and Wilson sent him on two more trips to Russia. Fairchild, who may have been jealous of Hansen's close relationship with his boss, accused Plant Explorer Number One of keeping bad records, overspending, and—perhaps an explorer's biggest sin—passing off plants he bought in a market as material he found in the wild. On December 9, 1908, when Hansen was in central Siberia, Fairchild reported that he had sent fourteen baskets of unidentified seeds. "We are in suspense as to the character of the large amount of material he has secured," he wrote in an official report about Hansen's trip.

After James Wilson left the department in March 1913, Fairchild's attitude made it hard for Hansen to get more plant exploring assignments from the federal government. "He had the gall to get his friends to write the new secretary, urging him to grant him $5,000 and carte blanche so that he would not have to take out any vouchers," Fairchild complained in spring 1913. "The present Secretary of Agriculture will handle this matter differently from the previous one."[6]

*   *   *

The department's second staff explorer, who was hired in July 1898, earned Fairchild's great respect. He was Mark Alfred Carleton, Fairchild's classmate at Kansas State Agricultural College, who had become a cereal specialist for the department after graduation. Carleton's great passion was to improve the grains cultivated in America's wheat belt.

Born in Ohio and raised on a farm in Kansas, Carleton spent his childhood and youth watching his neighbors labor constantly to harvest good wheat. Most wheat cultivated in America at this time was a red or white winter variety with soft kernels high in starch and low in protein. America's earliest settlers had planted it east of the Mississippi River and ground it into flour to make bread and pastry.

As pioneers moved west early in the nineteenth century, they brought seeds of these soft wheats with them, unaware that the varieties couldn't handle the different growing conditions west of the Mississippi. Midwestern winters are too cold and summers are too hot and dry for most soft wheats. In the prairie fields of Kansas, Carleton learned, they were especially vulnerable to rust, a fungus that shrivels the grain and rots the straw.

Carleton had also learned, however, that not all farmers in Kansas had this problem. The exceptions were Mennonites who had arrived from Russia in 1873. America was the most recent home for these Protestants, who had wandered through Europe for generations. The sect had originally lived in West Prussia, but many members moved to southern Russia about 1770 when Catherine the Great convinced them to settle remote sections of her country in exchange for one hundred years of special privileges, including exemption from military service. The Mennonites were skilled farmers who thrived in the Crimea by developing through trial and error hard wheat varieties that could handle the tough climate there.

In the mid-1800s, as Catherine's century of protection drew to an end, the Russian government warned the Mennonites that they would soon face conscription despite their pacifist convictions. Many in the community fled Russia and sought religious freedom in the New World. One of the first immigrants was Bernhard Warkentin, a Mennonite miller from Ukraine, who moved to Newton, Kansas in 1872. He arrived as America was looking for farmers to fill its vast open spaces. The Atchison, Topeka and Santa Fe Railroad, eager to settle the empty areas it owned along its first transcontinental route, recruited Mennonites—with Warkentin's help—to relocate from Russia to the Midwest. The company sold them land at the cheap prices of $2.50 to $5 an acre (well under $150 an acre in today's dollars) and chartered a steamer to transport their belongings from the shores of the Red Sea to New York. Trains carried thousands of people and their possessions to Nebraska and Kansas, where they lived in railroad company barracks until they picked the land they wanted. About 9,000 Mennonites settled in Kansas, most in the middle of the state near the towns of Newton, Halstead, and Moundridge. They felt at home there. "A traveler on the plains of Kansas, if suddenly transported while asleep to southern Russia and deposited in the Crimea, would discover very little difference in his surroundings," Carleton wrote in the 1915 *Yearbook of Agriculture.*

The newcomers' most valuable possessions were their seeds of hard winter wheat, varieties the Mennonites knew could thrive despite the Great Plains' rough growing conditions. In 1898, while Carleton was working at an experiment station in Lincoln, Nebraska, he saw how superior hard wheat was to soft wheat; he was certain that if other farmers in Kansas had enough good material, their yields would increase, too. After the government established the office of seed and plant introduction, Carleton pressured Fairchild and Wilson to send him out to get seeds. "I must go to Russia," he said.[7]

On July 4, 1898—only three days after Fairchild and Swingle's scheme provided twenty thousand dollars for plant hunters—Carleton left Washington for Siberia, then home to Mennonites who had relocated within Russia. Warkentin, the owner of a flourmill in Kansas, had given Carleton the names of people to contact. By late 1898 Carlton was in the Kirghiz Steppes in western Siberia, an area Fairchild called the Black Lands of Russia, buying hard wheat for Nebraska. U.S. officials' predictions of international famine made Carleton's trip vital. "These prognostications spurred our desire to find drought-resistant wheat in order to extend cereal production into the arid regions of our country," Fairchild explained.[8]

After exploring for six months, Carleton returned to Washington with several types of wheat, including the hardest of all—durum, often known as macaroni wheat. In a town near Orenburg he had bought six bushels of a durum strain called Kubanka from a Mr. Gnyozdilof.

While midwestern farmers were pleased with Carleton's seeds, midwestern millers were not. They didn't want the trouble and expense of updating their machinery to process harder grains. "Durum, the hardest of hard wheats, met at once with the most violent opposition, chiefly from millers, but also from all grain men," Carleton wrote later. "Various epithets, such as 'bastard' and 'goose,' were applied to the wheat without restriction."[9]

With Fairchild's help, however, Carleton campaigned to change the industry's attitude by advertising the superior taste of durum wheat. Carleton wrote a special bulletin filled with recipes for the new flour, including one irresistible to Americans: new and improved macaroni and cheese.

Before Carleton went to Russia, most Americans—at least those without Italian relatives—didn't know what good macaroni tasted like. "The most common form in which macaroni is served in this country is a very white, pasty, doughy mass of sticks, served in dilute tomato sauce," Carleton said. "The most enthusiastic lover of macaroni would have very little if anything to do with a dish of that kind."[10]

Carleton's promotional campaign worked. Within a few years, large grain processors relented and modified their mills to grind hard wheat into flour. Carleton's trip cost the U.S. government about $10,000 (about $250,000 today); by 1905 the new crop was worth $10 million a year (more than $250 million today)—a 1,000 percent increase. America had so much durum wheat that the country exported 6 million bushels a year. By 2011 production rose to about 50 million bushels a year. Because of Mark Carleton, American farmers had more than enough wheat, freeing experts at the end of the nineteenth century to worry about something other than widespread famine.

* * *

In September 1898, while Carleton was in Siberia, Fairchild met in Washington with the government's third plant explorer, Seaman A. Knapp. A lifelong friend and contemporary of Secretary Wilson, Knapp had been an architect of the agriculture department's system of state experiment stations. For most of his life, Knapp had lived in Iowa, where he farmed and taught; in 1886, at the age of fifty-three, he moved to Louisiana to set up a rice plantation. Soon after the office of seed and plant introduction opened for business, Wilson asked Knapp, who was sixty-five years old, to travel even farther from home and go to Japan to collect new rice varieties.

At the time, Americans consumed rice primarily as a pudding, not—like most people in the world—as part of a meal's main course. Americans demanded kernels with a clean, smooth texture. Farmers in Louisiana and Texas grew mostly long-grain varieties originally imported from Honduras, but the kernel's length made the rice fragile. When the outer coating was polished to whiten the grains, the only kind most Americans would eat, the rice often shattered. To make the product pretty and smooth enough to attract shoppers, processors coated it with paraffin wax. Of course, this beauty came with a price; buffing removed rice's nutrients and wax removed its taste.

Knapp's mission was so specialized and straightforward it sounded almost like a modern business trip. In Japan he researched the best kinds of short-grain rice and, in a straightforward deal, bought ten tons of it for about $18,000. Knapp shipped the rice to Louisiana and Texas, where farmers planted the seeds and, in only one growing season, transformed the South's rice industry. By 1905 the crop was worth almost $3 million (or more than $75 million in today's money).

Despite Knapp's improvement, America's rice-eating habits appalled Fairchild. "Rice is the greatest food staple in the world, more people living on it than on any other, and yet Americans know so little about it that they are actually throwing away the best part of the grains of rice and are eating only the tasteless, starchy, proteinless remainder," he wrote in a magazine article. He mocked Americans for demanding rice as shiny as "glass beads." "A pudding of stewed, sweetened rice, dusted with cinnamon is about as unappetizing to a fastidious Japanese as a sugar-coated beefsteak filled with raisins would be to an American," Fairchild wrote.[11]

Those glass beads were unhealthy as well. In 1908, a decade after Knapp's trip, scientists determined that a diet of polished white rice could cause beriberi, a discovery that forced rice growers to enrich the grains with the nutrients removed by milling.

*　*　*

Knapp's work—quick, straightforward, free of sentiment and, apparently, adventure—was the kind Wilson valued. Fairchild, the romantic, had more respect for the exploring done by his closest friend, Walter Swingle.

After he helped Fairchild establish the plant introduction office in August 1897, Swingle moved to Italy, partly to hunt for plants and partly to study botany. In the spring of 1898, he got a chance to do both. Swingle spent April reading scientific journals at the Zoological Station in Naples, the same place Fairchild had visited until Lathrop lured him away in 1894. The high windows of the station's narrow, book-lined library faced east

FIGURE 7. Walter Swingle working at his desk in Washington. By permission of the Fairchild Tropical Botanic Garden, 4850, David Fairchild collection.

across the Bay of Naples where plumes of smoke from Mt. Vesuvius wafted in the distance. Swingle, a handsome young man with a mischievous twinkle in his eye, spent days in the room trying to understand a fascinating but nettlesome subject: the sex life of figs.

Swingle was in the library because officials in Washington had promised to help American farmers cash in on a major turn-of-the-century food fad, Smyrna figs. Consumers spent about $400,000 a year—well over $10 million these days—importing the sweet, golden figs from Turkey. California farmers had tried for years to pocket some of that money by growing the same variety, but none had fruited properly.

In 1880 a California newspaper's promotional campaign increased the farmers' frustration. Guilian P. Rixford, a reporter who covered the San Francisco waterfront for the *Evening Bulletin*, was promoted to business manager for the paper. Determined to increase circulation, he imported

14,000 cuttings of Smyrna figs from Turkey and gave them away to subscribers.

Farmers throughout the state planted them and eagerly anticipated a bounty of fat, juicy figs. They were disappointed when the trees flourished but bore no fruit. Instead, they produced only desiccated figs the size of marbles that dropped to the ground before they ripened. For almost a decade, not one Smyrna fig grew to maturity in America. Some California farmers believed the newspaper had tricked them into accepting an inferior variety. Others were convinced the Turks had sent cuttings from sterile trees to protect their export trade. Tempers flared.

Humans have been passionate about figs since the beginnings of time. Although its true origins are unknown, the fig has one of the richest histories of any food. It has grown for thousands of years in Arabia, and at one time it flourished, along with olives, almonds, and dates, in the Garden of Eden. Some scholars believe the forbidden fruit of the Tree of Knowledge was a fig, not an apple. As Phoenicians colonized the Mediterranean basin and planted the trees throughout the known world, fig orchards spread through southern Europe and northern Africa.

Few loved figs more than the ancient Greeks and Romans, who honored the fruit in their myths and rituals. The Greeks, who considered the fig tree a sacred symbol of fertility, paid homage to the god Dionysus by presenting him with a vessel of wine and a basket of figs. Triumphant athletes were rewarded with a crown of fig branches. Later, at the beginning of the first millennium, Pliny the Younger wrote extensively about figs' health benefits, even asserting that eating them made old people look younger by reducing the wrinkles on their faces.

Like almost all fruits and vegetables now common in America, the fig—which is a member of the mulberry family—is not considered native to the New World. Early English colonists carried many fruit varieties with them, but records show the fig was one of the few that flourished. Captain John Smith, the Jamestown settlement's leader, wrote in 1629 that Mistress Jane Pierce, who he said was an honest and industrious woman, had gathered from her garden one hundred bushels of excellent "figges."

Other settlers determined that California's climate was better than Virginia's. Spanish and Portuguese missionaries brought fig trees from home in the eighteenth century. Most of their contributions have been lost, but one still thrives. In 1769 Junipero Serro, a Franciscan missionary from Mexico, planted the variety now called the Black Mission in San Diego.

For a century immigrants from Europe and settlers from the East Coast brought cuttings from home and raised fresh figs for their families. By the end of the nineteenth century, more than one hundred kinds grew in America. But none was as prized as one import: the Sari Lop ("yellow delicious" in Arabic) that grew in the sheltered valleys of the Meander River near Smyrna, Turkey. The figs there have been celebrated for more than 2,000 years. Swingle said it was probably the oldest continuing fruit industry in the world.

Of all the farmers in California who wanted to grow Smyrna figs, no one was as determined to succeed as George Roeding or Gustav Eisen, the owner and manager of the Fancher Creek Nursery in Fresno. It was a difficult effort, one so demanding that Roeding had inherited the fight from his father, Frederick, after he gave up. "At times the task appeared to be a hopeless one," Roeding said later, "and the temptation to dig up the orchard . . . was very great."[12]

The men believed the solution to the problem was a tiny insect—about half the size of the letter *w* on this page—called the *Blastophaga grossorum*. Locating this wasp, Roeding said later, "caused me so many sleepless nights, so many thousands of dollars, and such a tremendous effort."[13]

The flowers of all figs grow inside the fruit, inaccessible to pollinators like wind, birds, and bees. Most common figs perpetuate themselves because they are unisexual beings. The Smyrna variety, however, has only female flowers and needs to be fertilized by pollen from its natural male partner, the wild Capri fig. (Capri figs are inedible; their name derives from the Latin word for goat, a clue that the fruit tastes so bad that only a goat would eat it.) Roeding and Eisen understood a *Blastophaga* wasp from a Capri fig tree was essential to pollinate a Smyrna fig flower and allow the fruit to develop completely.

Ancient farmers recognized the importance of this pollination, a technique called caprification, which is one of the oldest plant breeding tricks in history. Herodotus described the process about 450 BC and Aristotle explained it more clearly 100 years later (*History of Animals*, book 5, chapter 26).

"The fruits of the Capri fig contain small animals called psenes," Aristotle wrote. "These are at first small grubs, and when their envelopes are broken, psenes, which fly, come out. They then enter the fruits of the fig tree and the punctures which they make there prevent these fruits from falling before they are ripe." As Swingle said later, Aristotle's account "could scarcely be improved today."[14]

In ancient and modern Greece, farmers hung Capri figs on the branches of Smyrna fig trees to encourage pollination.

Roeding proved, at least to himself, that the technique worked by meticulously carrying pollen by hand from his male Capri fig to his female Smyrna fig, first with a toothpick, next with a quill—which resulted in four ripe figs—and later with a blowpipe. By 1890 he had farmed 150 juicy Smyrna figs this way, but the method was far too slow and labor-intensive for commercial production.

Other California farmers didn't believe Eisen and Roeding, let alone Herodotus and Aristotle. Farmers sneered at the pair when they addressed local meetings to encourage the practice. "When I first announced my final conclusions about caprification and of the necessity of importing the *Blastophaga* at a horticultural meeting held in Fresno, 1887," Eisen wrote on October 3, 1939, "I was hooted down and some of the mob whistled."[15]

The farmers believed only soil, climate, and vicissitudes of the seasons determined the success or failure of a Smyrna fig tree. They considered caprification, in Swingle's words, "merely a peasant's superstition analogous to the hanging of horseshoes in favorite fruit trees to make them fertile."[16]

* * *

Walter Swingle, a brilliant young man who understood both biology and commercial farming, eagerly tackled the challenge of understanding Smyrna figs. Nothing excited him as much as a new scheme.

His intelligence had always been obvious. He was born on a farm in Pennsylvania, although his family moved to a bigger spread in Manhattan, Kansas, when he was two years old. Manhattan was a community of about 1,200 citizens who lived in sturdy stone houses, but when the Swingle family arrived, it was about to expand dramatically as railroad routes multiplied through the area.

When Swingle was five, his parents sent him to a one-room schoolhouse where students of all grades studied together. He took only four years to learn everything the instructors taught, so he dropped out when he was nine to help his parents on the family farm. He cut wood, shucked wheat, and picked corn with his father and churned butter and plucked chickens for his mother. She paid him 10 cents a churn so he could save enough money to subscribe to the *Youth's Companion*, a magazine that offered wholesome but tempting glimpses of the world beyond Kansas. When he was fifteen, Swingle enrolled in the state agricultural college.

As a boy Swingle loved plants, although he didn't know their formal names. He invented his own list, a personal taxonomy, which he used until he learned the scientific names in college. He taught himself German so he could read the original versions of important botanical articles. When he needed help translating, he consulted a German immigrant who fed the pigs at Kansas State's plant experiment station. As an undergraduate, Swingle assisted William A. Kellerman, the college's biology professor, who honored him by sharing the credit on a half dozen scientific articles on fungi that they prepared together.

In 1887, when Swingle was only sixteen, he delivered a lecture about plant diseases to Kansas State's science society. He was shy and stuttered terribly, but he had taught himself to speak carefully and softly to slow down his speech and avoid tripping over his words. Yet Swingle was so nervous about speaking in public that he brought his mother to the lecture for reassurance.

Fairchild, eighteen, attended the lecture and never forgot Swingle's performance. "It was an entirely new subject to us and we sat spellbound while he presented his discourse," Fairchild remembered. "Entranced, we watched Swingle's long arms wave about and his piercing gray eyes dart from one to the other of us." It was the first time Fairchild had heard the word *bacteria*.[17]

Erwin F. Smith was directing the U.S. Agriculture Department's division of plant pathology in 1889 when he visited the college and met Swingle, who was eighteen. Smith too was astonished by his intelligence. "I saw more or less of him for a week or two and set him down as one of the brightest young men I had ever met and felt that we must certainly have him in our division, if possible," Smith wrote on May 26, 1895.[18]

As soon as he could, Beverly Galloway added Swingle to his growing team in Washington. In 1891 Swingle joined Fairchild as a low-level government botanist trying to unravel the mysteries of the many plant diseases that threatened farmers' crops. Apples and grapes had problems with pests. Wheat was vulnerable to rusts and blights. Citrus fruits succumbed to sudden frosts and mysterious diseases. The department tried to help everyone, although citrus farmers were a special case because they had a powerful ally in Washington, D.C.

Ellen Barstow Platt, the wife of Thomas C. Platt, the Republican boss of New York State and a past and future U.S. senator, was a citrus grower in Florida. About the time the government hired Swingle, she complained to the secretary of agriculture that something mysterious was plaguing the oranges on her property. She asked for help immediately—and she got it.

Within weeks of joining the department, Swingle was on his way from Washington to Eustis, Florida—a twenty-nine-hour train trip—to deal with Mrs. Platt's problem. Because most oranges sold in America were imported from Spain and Sicily, Swingle had never seen an orange tree. When he arrived in Florida, he sent Fairchild a postcard describing them in terms a Kansan would understand. They are "something like oaks, but with orange-colored fruits hanging from their branches," Swingle wrote.[19]

He didn't know much about what oranges tasted like, either. One day in Florida he noticed a tree full of big reddish fruit and asked the owner if he could buy some. "Sonny," the owner answered, "you can have all those oranges you want. They won't cost you a cent."

Swingle filled a large pail with oranges, settled down in the shade, and started peeling them. He put half an orange in his mouth and panicked when the juice hit his tongue. "For a minute, I couldn't believe what was happening," he wrote later. "It burned and tasted like a cupful of pure acid." Swingle was too ignorant to know he had picked inedible fruit. "I knew I'd made a bad mistake and decided right then that I had a lot to learn about oranges, and I'd better get busy right away if I wanted to stay in business," he said later.[20]

Oranges kept Swingle busy for the next six years. He worked with Herbert Webber, his assistant who was six years older than he was, to design and build a laboratory to develop hybrids that could resist diseases. He worked on the project until the winter of 1894 when the record-breaking freeze hit and killed virtually every orange and lemon tree in Florida down to the ground. Swingle, who was as restless and eager as Fairchild to see the world, took a one-year leave from the government to study in Bonn, Germany, and Naples, Italy.

Meanwhile, George Roeding, the California fig grower who was seeking help from the agriculture department, had contacted Leland Howard, the department's chief entomologist. Roeding had raised almost 5,000 Smyrna trees and 95 Capri fig trees on his sixty-two-acre orchard, but he had no natural way to deliver pollen from one to the other. He needed wasps. Howard knew Swingle was both a scholar and a farmer, so he asked him to research fig wasps and explore the Mediterranean for *Blastophaga* while he was in Europe.

Swingle was delighted by the assignment. After he read all the literature he could find in Naples, he explored. On April 9, 1898, he hiked through fig groves in Resina, a medieval village at the foot of Vesuvius. The next day he visited another village called Portici on the Bay of Naples and the Chiaja neighborhood of the city of Naples; he also wandered through the fig groves at the Villa Sans Souci at Posilico, a walled estate owned by an English family.

Farmers in California were not the only people who didn't believe in caprification. In 1848, an Italian scientist named Guglielmo Gasparrini had published an article asserting that the wasps were useless. One day at the Naples Zoological Station, as Swingle was packing figs to send to Washington, two of the most eminent botanists in Europe visited him. One was Paul Mayer, an entomologist at the station; the other was Count Hermann zu Solms-Laubach, an authority on figs. The count, who was fifty-six, reprimanded Swingle, twenty-seven, for wasting time. "Why do you Americans spend good money to come to Europe to study things already decided?" he asked. "Gasparrini showed fifty years ago that figs do not need to be caprified to set fruit and that pollination has no beneficial effect. This is merely a peasant superstition. Why did you not . . . accomplish something useful and interesting?"[21]

Although he remained polite to men considered his betters, Swingle was not cowed. He stayed focused on his project and continued to collect figs. On his hikes he found a half dozen Capri figs he hoped would contain *Blastophaga* wasps. Because he had no shipping materials, he devised his own system: he wrapped each fig in thin foil from Italian cigarette packs and sealed the ends with wax, like a mustache.[22] Swingle cushioned the fruit in cotton and, on April 15, 1898, mailed a package about the size of two shoeboxes to Washington on the SS *Kaiser Wilhelm II*.

He complained about the expense to Fairchild, who was running the office in Washington. "This fig business has cost me a lot of money ($2 for cab fare on the 15th)," he wrote on April 19, 1898.

Nonetheless, Swingle was confident that his scholarly research, combined with his exploration of Italian fig orchards, would result in a breakthrough. "I have great hopes of the introduction succeeding, for a single insect should be enough to stock all of California," he told Fairchild eight days later.

Swingle's hopes were soon dashed, however. On May 25 Howard reported that almost all the figs had dried out by the time they arrived. Only one was in good shape and released hundreds of tiny wasps after it was opened in California. None, however, pollinated a Smyrna fig in Fresno. Swingle refused to give up. He had, as the writer of his *Miami Herald* obituary put it in 1952, "a mind like a bulldog. It never let go."

Through his research Swingle had learned that the Capri fig tree produces three crops a year. Only the third one, the fruit that remains on the tree over the winter, produces the wasp that can pollinate Smyrna figs. And that particular wasp won't work either unless it reaches the Smyrna fig at precisely the right moment.

In early spring 1899, Swingle set off again through the Mediterranean basin to collect hundreds of Capri figs growing under various conditions, hoping to find wasps at different stages of development. He wanted to increase the chance that at least a few would arrive in California at the right time to pollinate the Smyrna figs. He collected samples in Greece and worked his way north from Sicily up the Italian boot, from low-lying orchards to mountaintops.

In March Swingle took a ferry from Marseilles to Algeria to confer with Louis Trabut, a respected French botanist who ran a government experiment station in the province of Algiers. Trabut, who was also fascinated by the caprification process, generously gave Swingle more than four hundred Capri figs from his own garden. Swingle shipped six boxes of them to America.

Roeding expected this shipment to fail, too, yet he reported a month later that the figs had arrived in excellent condition. "They were quite firm, plump and green and looked as if they had just been picked," he wrote.[23]

Roeding opened several fruit and found them full of fig wasps on the verge of adulthood, almost ready for breeding. He put the open figs under one Capri fig tree and covered the tree with canvas to contain the wasps. Two months later, almost all the Capri figs had dropped to the ground, shriveled and useless—except twenty that were healthy and appeared to contain wasps. Roeding was elated.

That autumn Swingle, who had returned to the United States, went to California to see how his introductions were progressing. He was visiting Roeding at his nursery in Fresno on November 10, 1899, when he watched with delight as thousands of wasps emerged from Roeding's Capri figs, evidence that the wasps were finally living and breeding in California. In early June the next year, workers carefully strung individual Capri figs laden with wasps among Roeding's Smyrna trees.

On June 11, 1900, Roeding knew that his fig folly, as he called it, had finally paid off. He recognized that a single fig had been fertilized because its skin had lost its ribbed texture and swelled to become rounded and sleek.

On August 2, one healthy, fat, American-grown Smyrna fig fell to the ground in Fresno, California. For seven weeks, more ripe figs kept dropping. By the end of September, Roeding had six tons. "The introduction of the fig insect has been a great success," Swingle reported to his father in Kansas on November 2, 1900.

Roeding gave Swingle credit for the triumph. Because of his work, Roeding said later, "the insect finally decided, after so many years of constant and persistent work on my part, to become domiciled in the new world and, small as it was, to make history for the fig."[24]

Roeding soon expanded his production and modified the name of his fruit to "California Smyrna." He contracted that to "Calimyrna" after the name won a twenty-five dollar prize in his fig-naming contest. Today almost all the commercial figs grown in America come from California, and the most popular variety is Calimyrna, the result of Walter Swingle's intelligence and perseverance.

*　*　*

Shortly before Swingle left for Europe in 1898, Fairchild gave him a second assignment: date palms. Swingle later told a friend he had balked at the challenge. "We'll be old men before we can get a date industry going," he said he told Fairchild.[25]

Sweet dried dates were almost as popular in America as figs, although they too had to be imported from the Mediterranean region. The climate there was similar to parts of California and Arizona, but American farmers could not figure out how to produce enough fruit to establish a profitable business. Swingle set out to change that.

In March 1899, he crossed the Mediterranean from Marseilles to the north African coast and headed for another romantic destination: Biskra, an oasis of mud turrets and camels in Algeria that was the source of the world's best dates. "Biskra is a great place," Swingle wrote on March 20, 1899, to Orator Cook, another botanist in the department. "It is a grand sight to see thousands and tens of thousands of date palms standing out luxuriantly in absolutely sharp contrast to the utterly treeless desert!"

In Biskra he visited the office of a large French importer where he confidently marched up to a sales window and asked a busy salesman to explain date culture to him. Instead of helping, the salesman slammed the window "right in my face," Swingle said later. Left to figure out the business on his own, Swingle negotiated the best price he could for good cuttings. He paid 39 cents—a significant sum—for each one. "I had a great time bargaining with the Arabs for suckers. . . . I am irrigating the desert with money but not very many square miles of it," he wrote on March 20, 1899.

He packed the cuttings—which are called offshoots or suckers because they grow above ground—in five heavy wooden tubs partly filled with earth and shipped them to Tempe, Arizona, for transplant at an experiment garden. This time, however, the packing system he devised didn't work, and the plants arrived in bad shape.

About six months after the first batch died in Arizona, Swingle returned for more. This time he started in Paris where he behaved more correctly. He put on his top hat and frock coat and called at the headquarters of the largest date company in Algeria. The French man who ran the company agreed to help Swingle because he loved California. "When I was a young man my mother sent me around the world in the hope of building up my health, which at that time was bad and continued bad until I reached California," he told Swingle. "There I recovered rapidly the health which I have never lost since, and I am profoundly grateful to California for this priceless gift."[26]

A few weeks later, after Swingle had a clearer idea what he was doing, he traveled to Ourlana, a different oasis in the Sahara, to buy 405 cuttings of two date varieties, the Deglet Noor and the Rhars. Again Swingle packed the cuttings in heavy tubs for shipment on a steamship, but the captain balked at carrying them.

He "told me this was dangerous for such a large shipment because, in the case of a storm, tubs might break loose from their fastenings and roll all over the deck and do much damage," Swingle said later. "He would not carry them for less than 38 shillings, nearly $9.00 apiece. I could not possibly pay such a freight bill."[27] (That's about $250 each today.)

So Swingle, the young man who had mailed tiny figs in thin cigarette foil, invented a new system. Instead of heavy tubs, he wrapped the cuttings in damp moss, tied them with straw and banana leaves and packed them in old shoeboxes. He shipped the date palms that way for $9 a dozen, not $9 each.

Riding on a camel, Swingle accompanied his valuable cargo for two and a half days across the desert to Biskra. Next he switched to a special train and traveled three hundred miles to Algiers and finally sent the cuttings off alone on a ship to America. The date palm offshoots weighed eight tons, but the shipment was successful: about 80 percent survived the long trip. "Hurrah! for the success of your date palms," rejoiced Barbour Lathrop on October 30, 1900, after he learned that Swingle's experiments had worked. "I have faith in the future of that industry of which you may practically be called the father."

*　*　*

Figs and dates did not create the only romantic adventures in Swingle's life during his years abroad. While he was studying at the University of Leipzig in spring 1898, he hired a young Alsatian woman named Lucie Romsteadt to teach him French. The lessons went well. "I am now sure I can get French," he told Fairchild on July 14, 1898. "My teacher says I have a good pronunciation and I am learning a lot from her."

By August he told his father he was taking French lessons every day. And he told Fairchild about his new friend. "She seems a very refined girl of 25 or so and must have come from a family of some means, since she has studied music for years," Swingle wrote on August 7, 1898.

Lucie told Swingle she dreamed about moving to America to teach music and, within a few years, he helped that dream come true. They were married in Washington, D.C., on June 8, 1901, when Swingle was thirty.

Lucie Romsteadt Swingle was beautiful, charming and happy that she had married a handsome, brilliant, and, she thought, rich American man. Swingle, a scientist who didn't concern himself with money, adored his wife and bought her everything she liked. And she liked many things. "She wanted parties, dancing, clothes, fur, jewels, opening nights at the theater, and champagne suppers," Swingle told a friend years later.[28]

They were happy at the beginning of their marriage and their life together was filled with idyllic moments. When Swingle was on assignment in Europe in December 1902, for example, they visited Tuscany. Every day after he finished work, the couple rode bicycles together through the countryside. "The road we follow goes from here to Rome and follows the sea, often carried along steep cliffs by means of cuts and tunnels," he wrote his father from Livorno on December 5, 1902. "From this road we can see Elba where Napoleon had his villa until he escaped and, on particularly clear days, Corsica."

They biked past old castles and picnicked overlooking the rocky coast. But this romantic life was expensive, and Swingle soon realized that he needed extra money. At the end of 1902 he concocted a scheme to get rich by establishing the first date palm plantation in the United States.

By then Swingle knew more about date palms than anyone in America. He had learned how to purchase, package, and ship suckers to America. And, perhaps most important, he had valuable inside financial information: he was writing a definitive report on date palms that would disclose what he had learned. After the U.S. government issued his bulletin—the first one published in America—the demand for his new company's date palms and other services would increase dramatically.

"There would be a fortune in it and, once started, a princely revenue from the sale of the suckers alone," he told his father in his December 5 letter. Swingle calculated that the scheme would produce profits of about $150 (more than $4,000) an acre every year.

Swingle planned the business while he worked on his bulletin for the department. He arranged to get 160 acres with free water in Heber, a small town in the California desert, and buy 156 offshoots of Deglet Noor palms from Algeria. He even considered importing camels to work on the plantation.[29] He was, however, worried about competition. "The import business I have always counted an essential part of our scheme and we must not let it go out of our hands," he told his father, who was his partner on the deal, on March 19, 1904. "We have great advantage for the business, owing to our having gotten the start."

Swingle rushed to get the business ready by the time the government published his report. In the March 19 letter he implored his father to send another seventy-five dollars to Algeria for more plants. The bulletin, *The Date Palm and Its Utilization in the Southwestern States,* was finally issued on April 28, 1904. In it Swingle stated that the Salton Basin of California "is actually better adapted for this profitable culture than those parts of the Sahara Desert where the best export dates are produced."

Despite his knowledge and opportunity, Swingle's date scheme collapsed before he made any money. That spring the agriculture department was roiled by scandals about scientists in other divisions who allegedly profited from their research. There is no evidence that Swingle got into trouble, but by summer 1904 he had sold his shares in the company to avoid any conflict between his personal finances and his government job. He was disappointed and, because of Lucie's extravagance, broke. "I have the greatest confidence in date culture and am very sorry I had to sell my interests," he told his father, adding that he was anxious about money. "Have you been able to dispose of my shares?" he asked. "I should like very much to have six hundred dollars as soon as it is possible." His investment was worth about $16,000 in current dollars.

At this time Walter and Lucie Swingle were so deeply into debt that they sometimes didn't pay the rent on their Washington apartment. One time Swingle, facing imminent eviction, was forced to borrow money from his friends. They gave it to him, of course, and the Swingles went off together to pay their landlord. But on the way they spotted a new edition of the *Encyclopedia Britannica* in a shop window. They bought it immediately, paying for the books with the borrowed rent money. The next day Swingle had to return to his friends to ask for another loan.

The marriage was also troubled by Lucie's health problems. She had what was called "a delicate nature" and frequently caught various diseases. In 1909, as her health deteriorated dramatically, Swingle pleaded with Fairchild to give him a hundred dollars for her medical bills. On June 6, 1910, she died in Washington of typhoid fever. Lucie's death came as a relief to at least one of Swingle's friends, Marian Fairchild. "It was tragic rather than sad," she explained. "Lately her mind had been affected and [she] lived in terror of poisoning and all that sort of thing. Mr. Swingle is an attractive man as well as a brilliant one and his devotion to her has shut him off from all but a very few friends."[30]

Nine years of marriage to Lucie left Swingle with enormous debts. To pay them off, he gave up his own apartment and moved in with Orator Cook and his family. Within a year, Swingle turned his life around. In 1911 he met Maude Kellerman, the brainy and sensible daughter of his mentor at Kansas State. Swingle hired her to work for him as a botanist and, four years later, they married and stayed together for thirty-seven years until his death in 1952.

Romance in America

In the summer of 1903 David Fairchild was back in Washington, again imploring Secretary James Wilson to give him a job. This time, however, the entreaties worked. In September Wilson took him back, although at a salary so low—about $2,000 a year, an amount worth about $50,000 today—that Barbour Lathrop had another fit. "This putting you on the roll at the same pay you received four or five years ago is a clear evidence that the department refuses to recognize or even [make] a pretense of rewarding my loyalty and yours during the years you have passed in my employ," Lathrop complained on September 6, 1903, still steaming over the U.S. government's insufficient gratitude.

But Fairchild didn't grumble; he remained enthusiastic about his mission even when it didn't take him anywhere near a palm tree or a luscious fruit. He had returned from his travels with a driving passion, one his friends said later was the key to his success in life. "He had discovered in himself that wonderful never-failing source of enthusiasm for new plants, the tastes and smells of new fruits, the appreciation of new vegetables, of seeds and beautiful trees and blossoms, and of all the great green world of nature," Marjory Douglas explained. "It was an enthusiasm so contagious, persuasive and magnetic that it would draw hundreds and hundreds of people to see the world with new eyes, as he saw it, a tremendous fresh outpouring of his spirit. It would be the secret of his future. It would never leave him."[1]

Fairchild's first major task was to prepare the agency's tenth inventory of imports, the official tally of every seed and plant sent to the department by botanists around the world. Compiling this inventory was the important but dreary job of listing and explaining each of the 4,395 items received since September 1900. Because Wilson had never found a permanent replacement while Fairchild traveled with Lathrop, no one had tackled the job in more than three years. The backlog was huge.

It is easy to picture messy piles of envelopes and scrawled labels stacked in office corners around the department. Fairchild carefully sorted through all the material and, in the introduction he wrote to the inventory, graciously thanked the correspondents who had sent material. He reserved his most effusive gratitude, however, for his patron.

"The office considers itself extremely fortunate to have enlisted the cooperation of such a public-spirited man as Mr. Lathrop, who has conducted these various explorations almost entirely at his own expense, with no other idea than that of benefiting the American public," Fairchild wrote on April 18, 1904, in the inventory, which included 1,000 items Fairchild and Lathrop had collected since 1900.

While he was doing his new job, Fairchild began visiting the private Cosmos Club on Lafayette Square where important scientists and writers congregated. At the club one evening Fairchild met Gilbert Grosvenor, twenty-eight, the new editor of the National Geographic Society's monthly magazine.

Like his chance meeting with Lathrop on the *Fulda*, this accidental encounter transformed David Fairchild's life. And, at the same time, it expanded plant exploring from a little-known government program to a nationally recognized adventure. After this meeting, Fairchild's passion became a favorite cause of two of the most influential men in the nation's capital.

*   *   *

Since it was founded in 1878, the Cosmos Club's elegant, wood-paneled rooms had been the site of many meetings of Washington's intellectual leaders. On January 13, 1888, an especially significant session was held in its large assembly room, a dimly lit chamber lined with oil portraits of distinguished-looking men. That evening Gardiner Greene Hubbard, a wealthy investor who was Alexander Graham Bell's father-in-law, and thirty-two other men established the National Geographic Society to promote international understanding. The next year the new society launched its own magazine with the stated purpose of informing its two hundred members about the world and, in Bell's phrase, "all that is in it."

After more than a decade, however, the magazine was a commercial failure despite its lofty goal; its dense articles attracted only about 1,000 readers each month, and the operation was losing money. Volunteers produced each issue under the direction of John Hyde, the volunteer editor-in-chief whose paying job was chief statistician of the U.S. Department of Agriculture. After Gardiner Hubbard died in 1897, Bell became the society's president, and early in 1899 he decided to jazz up the magazine by hiring a vigorous full-time editor.

To find a smart young man to take over, Bell approached his friend Edwin A. Grosvenor, an American scholar who had been a college professor in Istanbul for many years. After Grosvenor returned to Washington, his family socialized often with the Bells, and the inventor had grown especially fond of Grosvenor's identical twin sons, Edwin and Gilbert.

The boys looked and dressed so much alike that virtually no one—including the brilliant Alexander Graham Bell—could tell them apart. In February 1899, when Grosvenor's sons were twenty-four, Bell offered the new job of full-time editor to either twin, whoever wanted it. The salary was $1,200 a year (about $33,000 now).

By then, Alec Bell and his wife had become prominent and wealthy members of Washington's intellectual elite with two attractive and unattached teenaged daughters. When Bell offered him the job, Gilbert Grosvenor was already in love with Elsie May, the older daughter. He immediately quit his job teaching school in Englewood, New Jersey, and accepted the post—his brother became a lawyer instead and never married—and began an intense courtship of his new boss' daughter. Family correspondence showed that, although Elsie liked Gilbert, her parents were not certain he was a good enough match for their daughter. "It will never do for her to marry a poor man and have to live in a small house, yet she seems drifting that way," Mabel Bell complained to her husband on May 12, 1899, in a letter from her home in Washington. "Gilbert is here quite often enough and, what is worse, she monopolizes him all the time."

The editing job gave Grosvenor many reasons to visit the Bells' house. Elsie's parents were soon impressed by his perseverance and devotion, but they were still uncertain about his financial prospects. "I do think Gilbert is fine, but how is he going to have sufficient [income] to support a wife for years, and meanwhile Elsie will be losing the best part of her life," Mabel Bell wrote to her husband on June 2, 1899. "I really think she ought to be married pretty soon, or she will settle into habits that will make it very difficult for her to adapt herself to life with anyone." Elsie was twenty-one.

The Bells solved the problem by convincing the society's trustees to promote Grosvenor from assistant editor to associate editor, sufficient justification to increase his annual salary to two thousand dollars, which was then enough money to support a family. Elsie finally accepted his proposal in August 1900, and they married two months later in London.

Bell's campaign to liven up the magazine and attract more readers began in earnest after Grosvenor became *National Geographic*'s top editor. Bell's most insistent demand was to increase the number of photographs in each issue. "More dynamical pictures, pictures of life and action—pictures that tell a story to be continued in our text!" Bell urged him on March 5, 1900.

Grosvenor heeded Bell's advice, but the magazine remained pretty dull for several years. Its transformation came by accident one evening in December 1904. Grosvenor was frantic because he suddenly realized he didn't have enough material to fill the next edition. "The printer was urgently demanding copy for 11 pages in the January issue," Grosvenor wrote later. "There is no tyranny so absolute as a printer's deadline, but I simply did not have a good manuscript available."[2]

Fumbling around his desk, he spotted a thick, unopened envelope. He unsealed it and out fell a wad of photographs of Lhasa, Tibet, a mysterious land that few westerners had visited. Grosvenor picked the best ones to fill the empty pages and rushed them to the printer, on time. This photographic glimpse of an unknown corner of the world was so popular that it revived the magazine's fortunes.

* * *

In 1903, when Grosvenor met Fairchild at the Cosmos Club, he was, as always, looking for dramatic photographs of life in exotic lands. Fairchild had taken hundreds during his travels, and as he chatted with Grosvenor, he described one unforgettable scene he had captured. In May 1901 he had gone to North Africa to find date palms. When he landed in Tunis, he noticed an astonishing spectacle: strolling through town were young women wearing yards of brilliantly colored silk and tall pointed hats. Each woman weighed about 300 pounds. "I simply could not turn my eyes away from them," Fairchild wrote later, "and frequently turned my Kodak toward them too, although they did not like it."[3]

Tunisians told Fairchild the women were Jews who had been fattened for marriage. The local custom was for a man to propose by presenting a large silver or gold ankle bracelet to the parents of the young woman of his choice. If the parents approved of him, they placed the anklet on her leg and fed their daughter a mixture of fenugreek seeds, milk, and honey until she weighed 250 or 300 pounds and her ankle was thick enough to fill the ring. Then the couple could marry.

Grosvenor was immediately interested in both the photographs and the photographer. He invited Fairchild to deliver a formal lecture to the geographic society members about his trip to Baghdad, then another unknown, mysterious land.

Grosvenor was also looking for something more than dramatic photographs for his magazine, however. As a member of the Bell family, he knew Elsie's parents were concerned about the future of their younger daughter, Marian, who was twenty-three and still unmarried.

* * *

By the end of 1903, Marian Hubbard Graham Bell was bored. Her only real interest was sculpture, an activity she took up intermittently. Born four years after her father patented his invention of the telephone, her life had always been comfortable, although her accomplished, formidable parents closely supervised everything she did.

Her mother had been completely deaf since she caught scarlet fever at the age of five. When Mabel was growing up, her wealthy parents, Gardiner and Gertrude Hubbard of Boston, refused to allow their daughter to be isolated from the world as most deaf people were at the time. When Mabel was fifteen, the Hubbards hired Alec Bell, a Scottish immigrant who was an instructor at Boston University, to teach her to communicate with others. Bell was a good teacher who was fascinated by human communication; his mother was deaf, and his father, Melville Bell, had invented a method to improve elocution called Visible Speech. (This invention probably inspired Melville Bell's friend George Bernard Shaw to create the character of Henry Higgins in *Pygmalion*.) After a few years, Alec Bell fell in love with Mabel Hubbard and stopped giving her lessons.

As a low-paid instructor and part-time inventor who experimented constantly with sound and speech, he couldn't afford to get married. For several years, Bell had had brilliant ideas, but he had never bothered to try to sell them. Before Gardiner Hubbard would allow his daughter to marry Alec Bell, he insisted that Bell obtain a patent on one invention that he thought had promising commercial applications. It was wired communication.

After the two men incorporated the Bell Telephone Company in July 1877 and sold most of its shares to the public, Alec Bell was rich enough to support Mabel Hubbard in the style her parents demanded.

Elsie May Bell, who was born a year later in London where her father was promoting his new telephone business, was their oldest child. Marian was born two years later on February 15, 1880, after the family moved to Washington, D.C. "My little girl weighed six and a half pounds, has long thick black hair," Mabel Bell wrote to her mother-in-law in an undated letter soon after Marian's birth. "She is a nice, soft, fat little thing."

Being the child of a celebrated inventor wasn't easy. From the beginning, Bell's many projects dominated Marian's life. Four days after her birth, Bell perfected another exciting creation: a wireless optical device called the photophone that transmitted sound by light. The two events were tied together in his mind for the rest of his life. "Only think," Bell wrote to his father at the end of February 1880. "Two babies in one week! . . . Both strong, vigorous, healthy young things, and both destined I trust to grow into something great in the future." Although Bell didn't fully develop this invention, scientists one hundred years later said the device was an early version of fiber-optic cable.

Bell was so proud of the invention that the family joked he wanted to name his new daughter "Photophone." Although she wound up with the name Marian, her relatives and close friends always called her Daisy.

Illness and death marked her childhood. When she was one year old, her mother gave birth to a son, Edward, who lived only a few hours. Bell was absent during the birth and quick death of his first son because he was away from home trying, unsuccessfully, to save President James Garfield's life from an assassin's bullet.

When Marian was three, her mother had a second son, Robert, but he too lived only a few hours. (Bell was away that time, also.) Mabel Bell's private letters show she felt guilty about failing to have a son to carry on the illustrious family name of Alexander Graham Bell. (It sounded illustrious by design. At birth his name was simply Alexander Bell, but when he was eleven, he added the middle name to sound more distinguished.)

Marian was close to her sister, who suffered from an extreme nervous condition during most of her childhood. Elsie had her first convulsion when she was about six years old and Marian was four. When Elsie was about twelve and her jerkiness and twitching worsened, she was diagnosed with a condition called St. Vitus' Dance. Her parents sent her to Philadelphia to live for a year with Dr. Weir Mitchell, a celebrated neurologist. Marian too was often sick with more common childhood diseases, sometimes for months at a time. "I had scarlet fever, diphtheria, and pneumonia all at once and survived," she said later.[4]

Despite continual cases of sickness and death in the family, letters show Daisy Bell was a cheerful child, determined to behave properly and make life pleasant for others, especially her deaf mother. "Her heart spontaneously moves her to help Mabel at every turn," Alec Bell wrote on November 10, 1892. "In a hundred little ways she tries to be of use—and yet all the time is delightfully unconscious of the fact. She has already become her mother's right hand."

Marian accepted her unusual family situation without complaint. Because Bell was preoccupied with his experiments, his wife and daughters often stayed alone in their Washington, D.C., home or traveled without him. At the end of 1891 Mabel Bell took both daughters on a long tour of Europe to celebrate Elsie's recovery from St. Vitus' Dance while Alec Bell stayed in America to work. The separation was difficult for the family, but Marian, eleven, remained cheerful. The three celebrated Christmas 1891 in their room in the Hotel de Genes in Genoa, Italy. Mabel Bell was depressed about the holiday, but Marian was content.

"She is a happy-natured little thing and her bright face is my great comfort now," Mabel Bell wrote to her husband that day. "Daisy insisted on having a Christmas tree of her own. It was a pathetic little attempt, a tall

FIGURE 8. Marian Bell, whose nickname was Daisy, seated in front of her sister, Elsie, and their parents, Alexander Graham Bell and Mabel Hubbard Bell. Photo from the U.S. Library of Congress, item no. 00649975.

laurel branch set in our water pitcher. . . . Daisy has announced that she has had 'a lovely time today' and that her tree 'was almost the nicest she ever had.' I am perfectly astounded. Does it really take so little to make a child happy?" Her parents were proud of her independence and good nature. "Daisy especially, I think, is going to develop into a self-reliant and beautiful woman," Alec Bell wrote when Marian was 12.[5]

Although the sisters were close friends, they were very different. Elsie was gregarious, lively, and eager to be the center of attention. Marian was shy, serious, and generous. "She 'carries her ain sunshine wi her' wherever she goes, and the house is dark when she goes out," Alec Bell, who never lost his love for his native Scotland, told his mother in a letter.[6]

The Bells owned a mansion at 1331 Connecticut Avenue near Dupont Circle in Washington and a large house in the mountains at Baddeck on Cape Breton in Nova Scotia. Their country estate, which they called Beinn Bhreagh ("beautiful mountain" in Gaelic), was a spectacular site that Alec and Mabel Bell discovered accidentally when they were briefly shipwrecked nearby. For years they bought all the land they could assemble until their property covered six hundred acres.

Although they lived well, the Bells often ran short of money. In the early years of their marriage, they sold the majority of their telephone company shares and devoted much of their remaining income to helping deaf people, financing Alec Bell's experiments, and promoting the *National Geographic Magazine* and other scientific journals. Marian was not raised to think of herself as an heiress. Although her parents gave her an allowance of three dollars (about seventy-five dollars now) when she was fourteen, she was required to explain exactly how she spent the money each week.

During Marian's childhood, Alec Bell moved to Nova Scotia every summer because he hated the heat and humidity in Washington. The two girls frolicked happily there as their mother flouted convention and dressed them in boys' clothes so they could run freely. As Marian and Elsie matured, they spent time in Europe with their mother studying French, art, and music, all essential skills for young women at the turn of the century. Charles Thompson, Alec Bell's personal assistant, sometimes escorted them.

During a trip to Rome in 1892, when the girls were twelve and fourteen, Mabel Bell told her husband each had decided what she wanted to do when she grew up. Their dreams were prophetic. "Daisy thinks she wants to go sightseeing all the time," Mabel wrote on February 26, 1892. "Elsie thinks she would like to visit strange countries, but only where she can go into society."

In May 1895 Mabel Bell left her husband in Nova Scotia and took the girls to France to study language. Despite the family's comparative wealth and considerable fame, correspondence shows that Mabel worried about money. "So far I have simply failed, miserably and disgracefully, and I am down," she told her husband on May 16, 1895, from her room at the Hotel de Vendome in Paris. "I have lain awake nights—I, who am such a good sleeper—going over my accounts over and over, trying to see just how much I may spend. I have come to the conclusion that another week of this hotel will simply ruin me and I am frantic to leave, but until today it just seemed as if I simply couldn't find a hole or corner in Paris where there weren't English or American people. I got so tired and discouraged." That day Mabel, Elsie, and Marian Bell ate lunch in a restaurant for two francs rather than spend six francs to eat at their hotel.

After Mabel's sleepless night, the family moved into a convent in Paris where costs were low and French speakers abounded. Mabel Bell also took her daughters to the Louvre Museum to see sculpture, an excursion that may have triggered Marian's lifelong love of the art form. "We spent an hour among the naked statues," she told her husband on June 2, 1895. "Both were vastly edified thereby and somewhat to my horror improved their knowledge of natural history. . . . It seems to me there used to be fig leaves about and there are none anymore."

Alec Bell's many projects included helping Helen Keller, the deaf mute girl who was Marian's age. After a month in the convent, Marian displayed a rare breach in her good nature when she complained about her father's absence. "Mamma, do you know that I don't think Papa quite appreciates us," she said. Mabel Bell told her husband about the comment, adding that his daughters were jealous of the attention he paid to Keller. Moreover, she said she agreed with them.

"I cannot think it unjustified," Mabel Bell told him on June 25, 1895. "I tell them to work hard and show you what they can do." These poignant reports affected Alec Bell. In his May 20 answer, he wrote, "Is it too late for me to show you that I really do love my wife and children above my experiments and work?" That summer he put aside his inventions long enough to join his family in Europe.

Mabel Bell worried about other matters too, especially what would become of her daughters because they had unconventional parents. Their father was an inventor who spent most of his time away from them, and their mother was deaf, a condition she said made her only "half a mother."[7]

She picked schools with teachers who educated her daughters while protecting them from their father's celebrity. For several years Marian went to Mount Vernon Academy, a progressive school for young women a few blocks from their home in Washington. There she became friends with Alice Barton Hill, a student four years older than she was. Alice had lived in Germany for several years and shared Marian's passion for art and music.

After her summer in the French convent, Marian switched to Miss Piatt's, a boarding school in Utica, New York, where her mother hoped she would make friends with down-to-earth people. One Thanksgiving Marian's mother encouraged her to spend the holiday with a school friend in Buffalo. "The people you meet in New York, Boston, Washington and a fashionable place like Morris Plains [New Jersey] are not representative Americans," she wrote on November 4, 1897. "I would think the citizens of Buffalo would be much more so, because Buffalo, while quite a large city, is not a fashionable one. It is more the home of respectable well-to-do cultivated people, the very best class of people."

Mabel Bell raised her daughters to be independent and broad-minded. In an undated letter when Marian was about seventeen, her mother explained her philosophy: "There is a feeling among a large number of poor people that rich people are too rich and have too many privileges, and they are trying all the time to lay taxes on property which will injure it very much. So I feel that it is every bit as important for you and Elsie to be able to support yourselves as if you were poor girls with nothing but a good education for inheritance."

In the spring of 1898 she told Elsie her own aspirations for them. "I want you to have a broad outlook on life," she wrote, "to know not only the life of your family circle and equals, but the life of other people—of the poor, of the army and navy, of students, of thinkers, of socialists as well as philosophers."

After finishing high school, Marian struggled to find her place in the world. Raised in privilege—although money had sometimes been tight—she had traveled with her family through Europe and Japan, learned to speak French, converse with strangers, and play the piano to accompany her sister. She was too intelligent to be totally satisfied by these activities, but she apparently believed she had few options. Family letters never mentioned college.

One intriguing opportunity presented itself in 1898 when Clara Barton, who was already a celebrated nurse, invited Marian to go with her to Cuba or Key West to treat soldiers injured in the Spanish-American War. Her mother thought it was a great chance for Marian to learn about the world, but her father vetoed the plan immediately. "At first I was inclined to think it a joke . . . [but] to treat as ridiculous a serious thought of a loving heart would only wound Daisy and accomplish no good," he wrote on June 8, 1898. "I am proud that my little girl has spirit enough and heart enough to want to go to the rescue of the suffering. It is just like her, always wanting to help others, even at her own expense." He said he refused to approve the trip because she could catch yellow fever.

After Elsie married Gilbert Grosvenor, Marian drifted more. She returned to Europe with her mother and spent her twenty-first birthday in Paris; again her father was too busy to mark the milestone properly. "I have been hoping that every ring of the doorbell might be a cable from you," Marian wrote him on her birthday.

None came, but a short note from Alec Bell arrived a few days later. He mentioned that his photophone invention was twenty-one years old, too, still giving his two creations equal attention. Mabel Bell reminded her husband how well Marian had turned out: she was happy, gentle, intelligent, and attractive. "The more you know her, the more she steals your heart," she told him on February 15, 1901.

Alec Bell soon shared his wife's uneasiness about Marian's future. On July 22, 1901, he told his daughter he was going to have his secretary type up the letters she wrote him from Europe. "You may also like to show your children and grandchildren an account of your adventures written at the time," he wrote. "So hurry up and have some adventures."

Since childhood, Marian had shown the deepest, steadiest interest in one subject: sculpture. During trips to Paris with her mother and sister, she had taken lessons and modeled alone in clay, even when her hotel room was so cold her fingers practically froze.

She had a few artistic mishaps along the way. In the fall of 1901, when she was living with her father in Nova Scotia, she tried to make a life mask of Susie McCurdy, a young family friend. After covering Susie's face with a protective coating, Marian applied several layers of plaster. As the plaster dried and tightened her skin, Susie's face began to hurt. Soon she had trouble breathing. Marian, frightened, didn't know what to do and ran to her father for help.

"When I arrived on the scene there was Susie on her back on a sofa—helpless—with a mass of plaster weighing several pounds completely covering her face and apparently as hard as stone," Alec Bell told his wife on November 17, 1901. The great inventor spent five and a half hours carefully chipping away the plaster with a meat cleaver. Then he gave Susie a shot of liquor and sent her to bed, uninjured.

When Elsie came of age, she had a debutante party at her parent's house in Washington. But Marian chose a different way to enter society: she took sculpture lessons. In the spring of 1904 she and Alice Hill, her friend from Mount Vernon Academy, left Washington for New York to study art, a thrilling and daring move for two unescorted women in their twenties. Probably because her father had vetoed her plan to help Clara Barton, Marian didn't tell him about New York until it was too late to stop her. He wasn't happy about the plan when he found out.

"I love my little girl very dearly and would like to help her in any way I can, but I am not given a chance," he complained to his wife on May 2, 1904. "Of course I am absorbed in my work, but it seems to me that in this whole matter Daisy has acted upon the tacit assumption that her father is not interested in her or in her work and that therefore there is no use in consulting me." Marian and her mother ignored his objections.

If the Malay archipelago was an exciting place for a young botanist early in the twentieth century, New York City was just as thrilling to young sculptors.

Marian and Alice had arranged to study for a month under Gutzon Borglum. At this time Borglum's career was just beginning, but he became a national celebrity after he created the presidential monument at Mount Rushmore. Borglum was iconoclastic, a character trait that appealed to Marian Bell. His studio was an old barn behind several brownstones on 38th Street near Third Avenue. Marian and Alice studied there for six weeks in the spring of 1904 and returned for a second session in the fall.

At first Marian was an uncertain student. Borglum "told Daisy that what she had to do was to get her own models and things and work out her own conceptions and he would help her with the technique," Mabel Bell reported to her husband after she visited Borglum's studio on May 2. "She said the trouble with her was she had no ideas. He replied she must get them."

Borglum was an activist artist who was undoubtedly pleased to have as students two innocent young women, one of them a famous inventor's daughter. When he wasn't giving advice, he assigned them mundane tasks around his studio. "Gutzon probably enjoyed making the pair of them sweep up the studio and mix clay," Borglum's widow wrote later. "Once he made them dye untold yards of muslin until their hands were completely discolored. 'Art was difficult,' he told them."[8]

Borglum may not have been a great teacher—he later told another assistant, Isamu Noguchi, he would never be a good sculptor—but Marian and Alice loved living as young bohemians in the big city. "Daisy is so pleased and happy at the idea of going to work in such inspiring surroundings and just wants to be allowed to throw herself wholly into the work for the month she is to be here, to have nothing to take her mind off it," her mother told Alec Bell in the May 2 letter.

The two women met many other young artists in New York City. They became friends with Jo Davidson, a sculptor who lived in Greenwich Village, and Clarence Dean, an architecture student at Columbia University. Shortly after they arrived in New York, Marian met a handsome young composer named Franklin Harris who had recently moved to Manhattan from Missouri Valley, Iowa. Marian immediately introduced him to Alice. When the two young women returned to New York in the fall of 1904, they shocked Mabel Bell—Marian certainly would not have mentioned the matter to her father—by attending a musical performance in his rooms on West 103rd Street without a chaperone. "I do not like the idea of your going to Mr. Harris' boarding house," she scolded Marian on October 24, 1904. "I do not think it is the thing at all. I cannot understand your doing it or Alice, either. I am very sorry."

Her disapproval apparently evaporated after Harris proposed to Alice and the Bell family accepted him, despite his bohemian ways. "I think on the whole I like him," Mabel Bell told her husband on November 8, 1904. "He has a fine face."

Elsie Bell met Harris for the first time a week later and told her father he was "not so eccentric and long-haired as I had expected. I'm glad, though, he isn't to be my brother-in-law!"

Unfortunately for Marian, her matchmaking backfired. After Alice Hill became engaged, she dropped out of the sculpture classes to prepare for her wedding, a decision that meant Marian would soon be left alone in New York City. In early November 1904, while Marian was trying to decide how to manage in the city without her friend, she was stricken with acute appendicitis. Although Marian recovered quickly, doctors said she risked another attack at any time. Under the circumstances, Mabel Bell refused to let Marian live alone, although she understood she was disconsolate about the abrupt end to her independence. "It seems to her that there is nothing ahead since Alice Hill is leaving her," Mabel Bell reported to her husband on November 11 after the health crisis passed. Marian, shaken by the sudden changes in her life, returned to Washington and her previous quiet routine at her parents' home. She was unhappy.

Mabel Bell pleaded with her husband to find interesting work for Marian in Washington. "I wish you could write Daisy and give her something to do for you," she wrote on November 21. "It is very hard for her being here just now, not quite well or strong and bitterly regretting the life in New York and missing her friend."

Marian had one serious beau, a young Washington lawyer named Robert Tyler who often served as her escort. As Gilbert Grosvenor had been with Elsie, Tyler was persistent; he wooed Marian for at least four years. At first her mother liked Tyler, but her letters show she worried that Marian would marry him out of sheer boredom. "He is a fine fellow, I doubt not, but I don't want him to have our Daisy and he will take her if he possibly can," Mabel Bell told her husband on May 1, 1901. "And the danger lies in that she knows nobody else. I wish she could be thrown more with other men."

Only a short while earlier, Gilbert Grosvenor had invited David Fairchild to speak about Baghdad to the National Geographic Society, including its most prominent member, Alexander Graham Bell. Bell apparently approved of Fairchild's speech because after the lecture he invited him to his home one Wednesday evening, when he hosted a regular salon for a revolving group of about twenty-five accomplished men in Washington. An invitation to the gathering, which was held regularly from autumn to spring, was considered an honor. "In those days," said Knowles Ryerson, a California botanist who later joined the agriculture department, "Bell's home was near what a European salon would be, an intellectual center with leading politicians, scientists, and writers congregating in his home."[9]

From their first meeting Fairchild was enormously impressed by the inventor. "Mr. Bell was at his charming best on these occasions, for he enjoyed his guests, drawing them out with courteous and interesting questions," Fairchild recalled. "He always made you feel that there was so much of interest in the universe, so many fascinating things to observe and to think about, that it was a criminal waste of time to indulge in gossip or trivial discussion."[10]

Meeting the famous Alexander Graham Bell was another lucky break for Fairchild, although the encounter was probably no accident. Fairchild most likely believed his exploring adventures alone warranted the invitation, but the Bell family correspondence suggests his status as a young, eligible bachelor had much to do with their attentions.

Fairchild wasn't eligible by choice. He had briefly been engaged to Mildred Howells, the daughter of another prominent American. Fairchild had met her and her father, William Dean Howells, the editor and writer, at Florence Lathrop Page's summer house in Maine in 1898. Fairchild, always the romantic, had fallen in love immediately. "From the very first time I set eyes on her, I cannot get her face out of my mind," he wrote his best friend on November 7, 1898. "It is awful, Swingle, but she haunts me."

Mildred was twenty-six years old when Fairchild met her. Talented and serious, she had already published several poems and exhibited her watercolors at an art show in Paris. James McNeil Whistler, another famous friend of her family, had painted her portrait.

Fairchild managed to see Mildred briefly before he set sail with Lathrop on their second expedition. "Think how awful on the eve of a two or more years' cruise to run across this creature and have her turn your head," he told Swingle in November 1898.

In October 1902, after Fairchild and Lathrop returned to America, Fairchild visited New York City to propose to Mildred, even though he was preparing to leave again soon. To his delight, she accepted. "You will be interested to hear that Miss Mildred Howells has consented to look upon me as her fiancé," Fairchild told Swingle in December 1902. "She is a beautiful character and has a host of friends who shower me with congratulations."

The *New York Times* published a short engagement announcement, erroneously describing Fairchild as an entomologist with the Smithsonian Institution rather than a botanist with the Department of Agriculture.

Henry James, another family friend, sent his good wishes to Mildred's father. "I wish her and you and the florally-minded young man (he must be a good 'un) all joy in the connection," James wrote on December 11, 1902.

The connection didn't last long, however. Only two months later, when Fairchild was sailing with Lathrop near Zanzibar, he received a letter from Mildred breaking off the engagement. Her reasons and Fairchild's reactions are unknown, but her family situation was complicated. Her sister had died in 1889, her brother worked as an architect, and her mother was an invalid, and so Mildred had become her father's essential helper. When William Dean Howells informed his relatives that the marriage was off, he explained she had a simple reason. "There is nothing against Mr. F.," a relative reported. "She simply changed her mind."[11]

Mildred Howells, who died at the age of ninety-three, spent her life caring for her father and his literary legacy. She never married.

* * *

A few weeks after Fairchild attended Bell's salon, where he apparently passed the family's inspection, Gilbert and Elsie Grosvenor invited him to dinner at their Washington home. It was a small party and Elsie Grosvenor assigned Fairchild to the seat next to her sister, who had returned from New York only a few days earlier. Marian was a graceful young woman of twenty-three with beautiful dark eyes and an independent spirit; all she wanted to talk about was sculpture. "I was fascinated by her," Fairchild recalled, although he assumed she was engaged to marry someone else.

Years of travel with Barbour Lathrop had taught Fairchild how to chat intelligently at dinner parties, so he managed to charm Marian, too. "Our conversation was largely on art, about which I knew nothing but could talk a good deal, having traveled with Mr. Lathrop, who was a real connoisseur," Fairchild wrote later.[12]

Marian Bell, longing for romance and adventure, was attracted immediately to the handsome scientist who had traveled twice around the world. The young couple saw each other frequently at holiday parties, concerts, and lectures in Washington through the end of 1904.

Many years later she explained that she was struck most by his personality, not his appearance. "He had such great personal *charm* and was always ready to be interested in other people and what they were doing," she wrote in an undated memo on file at the Fairchild Tropical Botanic Garden. "People wanted to have him around!"

Early in February 1905 Bell, who was a trustee of the Smithsonian Institution, finally helped Marian put her artistic skills to work. He brought her along when he and other trustees went to Detroit to evaluate a gift from Charles Lang Freer. Freer, a railroad car manufacturer, wanted to donate his vast collection of American and Asian art to the Smithsonian, but the trustees were not especially interested in the offer. Yet Marian's enthusiasm was strong enough to persuade the older, distinctly non-artistic trustees to accept Freer's collection. She was always proud of her role in the acquisition, which led to the establishment of the Freer Gallery of Art in Washington, D.C.

Soon after Marian returned to Washington from Detroit, Fairchild proposed. By March 1, 1905, they were engaged; six weeks later, they were married. It was an unorthodox, brief engagement, but it was typical of Marian. She was an early campaigner for women's suffrage and later became the first woman licensed to drive a car in the District of Columbia.[13] Their son later joked in an interview that Marian Bell agreed to marry David Fairchild only because he promised to take her to Java.[14]

Flying Machine Crank

The marriage of an unknown government botanist to the lively daughter of one of America's most prominent men caused a small newspaper sensation. The *New York Times* article about the match played up the slightly off-beat angle that Fairchild had given her an engagement necklace instead of an engagement ring. The *Philadelphia Press* reported, without any attribution, that Marian Bell had a dowry of $1 million—$24 million in current dollars—a statement that was probably based on rumor, not fact.

Even Agriculture Secretary James Wilson, who remained dubious about the value of Fairchild's international plant exploring scheme, was proud of him. "I congratulate you very cordially," he told Fairchild on March 10, 1905, after the engagement was announced. "It is something to go into the Bell family, as I regard him as our greatest scientist.... We take a great deal of interest in the development of our young men here and are glad to know that you have done so well in the most important step of your life."

The marriage instantly gave Fairchild and his passion for foreign foods two powerful allies in Bell and Grosvenor, men with far more influence in Washington than Barbour Lathrop. While Alec Bell may have been an inattentive father while his daughters were young, Mabel made certain he paid careful attention to them as adults. And that duty often meant advancing their husbands' careers. When Elsie was considering marrying Grosvenor, Mabel warned Bell that she would hold him responsible for their future son-in-law's success. "If Elsie decides she wants him, you will have to look out for him," she wrote on May 20, 1899, shortly before the couple became engaged.

Bell had done as he was told. He had protected and nurtured Grosvenor's rise at the magazine, and so, after Marian's wedding, it was Fairchild's turn. Family correspondence shows that Mabel Bell rarely passed up a chance to help him. "Daisy feels you don't realize the magnitude of David's work," Mabel wrote to Alec in Nova Scotia in November 1909. "I wish you would go to his office first thing when you come home."

Bell aided Fairchild in ways large and small. Although he was a hearty eater without any apparent interest in exotic dishes, Bell graciously served the plant explorers' unfamiliar foods to his important friends at his Wednesday evening salons and, once a year, at the National Geographic Society's banquet. The occasions were the first chance policy makers had to taste foods grown from seeds collected by Fairchild's plant hunters.

In the ballroom of the Willard Hotel, scientists, world leaders, and other banquet guests dined on dasheens and chayotes (both vegetables from the West Indies), jujubes from China, and the Deglet Noor dates Swingle had transplanted to California. Fairchild was delighted about this opportunity, which was valuable free advertising for his little office. Because the supply of foods was limited, however, Fairchild found it difficult to capitalize on whatever interest Bell's support generated.

Bell offered other career advice. He urged Fairchild to establish a Plant Utilization Society to educate Americans about the benefits of foreign foods (although he wasn't enthusiastic enough to bankroll the operation) and suggested specific ways to curry favor with his boss. "Secretary Wilson appreciates publicity of the right kind and it cannot be bad for David for the secretary to be favorably impressed by his public writings," Bell told his wife in a July 25, 1906, letter. "I think the secretary would be more inclined to read these reprints if they were sent to his private address instead of to the department where of course he is deluged with printed matter."

Suggestions like this may have been helpful, but they were minor compared to the inventor's greatest gift to his son-in-law: his attitude. Bell introduced Fairchild to a world that was bigger and more exciting than any place he had visited with Barbour Lathrop. Under Bell's influence Fairchild soon saw his foreign seeds and plants as more than food and crops. Bell taught Fairchild to consider his imports as players in the social and economic transformation under way at the beginning of the twentieth century.

Early in his career Fairchild had chronicled his travels in scientific journals and agriculture department bulletins, but he had never pursued a general audience until he met the Bell family. Ten months after the wedding, Grosvenor again invited Fairchild to speak to the National Geographic Society about his work. As Fairchild discussed the upcoming opportunity with Bell, the inventor prodded him to broaden his point of view. Fairchild's office didn't just traffic in seeds and plants, Bell asserted; it introduced new material that improved American food when it was combined with native plants. His work was groundbreaking activity that needed an impressive label. So when Fairchild delivered this key speech on February 9, 1906, he used a lofty label that Bell had suggested: "Our plant immigrants."

* * *

Bell's thinking about his own work had by then become even loftier. When Fairchild met him, Bell was preoccupied with solving the biggest scientific challenge of the day: human flight. He became so obsessed that he sometimes referred to himself as "the flying machine crank."[1] His experiments gave Fairchild a close-up view of the thrilling successes and crushing failures of the invention of the airplane. And it allowed him to watch as aviation made the world a much smaller place, transforming exploration completely.

Bell's family supported the inventor's fixation with flight. As early as June 5, 1893, Mabel told him, "I am very much interested indeed in your flying machines. At last you have come up with something I can understand."

Each of America's aviation pioneers had a different idea about how an aircraft could be built. Samuel Pierpont Langley, an astronomer who was the secretary of the Smithsonian Institution, favored two fixed wings next to each other, for example, and Orville and Wilbur Wright worked with gliders and then two fixed wings on top of each other. Bell preferred kites. These, he believed, combined strength and lightness and could land more safely than steam-powered machines. He didn't care if other experimenters scoffed at his idea. "The word 'kite' unfortunately is suggestive to most minds of a toy, just as the telephone at first was thought to be a toy," Bell told the Washington Academy of Sciences in 1907.

Bell was so determined to conquer the skies that he systematically conducted 1,200 tests over nineteen years at his estate in Nova Scotia. "Alec thinks these are the hardest, slowest, more tiresome and most unsatisfactory experiments he has ever attempted and the most discouraging," Mabel Bell told her mother in November 1897, "but he will not give up until he has got some results."[2]

During his summers in Nova Scotia, Bell had become close to two local young men. One, Douglas McCurdy, was the son of a Baddeck businessman who worked as Bell's nonscientific assistant for many years. The other, Frederick Walker Baldwin, who was called Casey, was a friend of McCurdy from the University of Toronto. Casey Baldwin was a civil engineering student who became interested in aviation after reading books written by Samuel Langley. Early in 1906 Baldwin proposed writing his senior thesis on flying, but the plan shocked his teachers. "Baldwin was summoned before the dean, who intimated kindly but firmly that a thesis on some more stable subject would be more acceptable to the faculty," Catherine MacKenzie, Bell's longtime secretary, wrote later.[3] In the summer of 1906, Bell hired Baldwin and McCurdy—whom the Bells always called "the boys"—to help him experiment with large kites.

David Fairchild was a guest at Bell's home the next summer when a new flying machine enthusiast joined the group. He was Thomas Etholen Selfridge of San Francisco, a 1903 graduate of the U.S. Military Academy at West Point and a lieutenant in the U.S. Signal Corps' aeronautical division in Fort Myer, Virginia. Selfridge was tall and slim and often wore old corduroy pants and flannel shirts when he wasn't in uniform. He fit easily into the Bell household and soon became Fairchild's good friend, later visiting him at his home in Maryland and once bringing Orville Wright with him. "He had acquired quite a lot of book knowledge [about aviation] but appeared a bit hazy on the subject in general from a practical standpoint," McCurdy wrote later about his first meeting with Selfridge. "In a few days, he was quite one of us and we all felt very much attached to him. He was a jolly fellow, could take a joke well and could handle the billiard cue to perfection."[4]

A few weeks after Selfridge arrived in Nova Scotia, Bell recruited another man to join the group. He was Glenn Curtiss, an inventor of

FIGURE 9. The Aerial Experiment Association in 1907. From left to right: Glenn Curtiss, Casey Baldwin, Alec Bell, Tom Selfridge, and Douglas McCurdy. John Alexander Douglas McCurdy/National Geographic Creative, image ID 599613.

motorcycle engines from Hammondsport, a town in the Finger Lakes region of upstate New York. Bell asked Curtiss to develop a lightweight engine that could power a large kite. After Curtiss arrived at his estate, Bell had his team in place. "Mr. Bell and the four brilliant young men were such a happy combination of personalities that Mrs. Bell could not bear to have the association end with the summer," Fairchild wrote later.[5]

At Mabel Bell's urging, they formed a partnership called the Aerial Experiment Association on October 1, 1907. The group had one simple goal that was, in Selfridge's phrase, "to get into the air by any means we could." The group spent a year designing and building different models. Starting with Bell's kites, their first three variations actually did fly, each trip a little longer and stronger than the last. And then came Curtiss's turn on July 4, 1908.

His design was a biplane with movable wing tips called the *June Bug*. The Aerial Experiment Association members were so confident that Curtiss's invention would succeed that they entered it, untested, in a national competition sponsored by the *Scientific American*. The magazine challenged inventors to build a manned, heavier-than-air flying machine that could stay aloft for one kilometer.

The Wright brothers had already flown their plane, called the *Flyer*, at Kitty Hawk, North Carolina, on December 17, 1903. But because only a handful of spectators had witnessed the event, most Americans didn't believe human flight was possible. At the time, Fairchild wrote in his autobiography, "Flying verged on the supernatural."

Unlike the Wright Brothers, Curtiss and the other members of Bell's group invited the public to watch the *June Bug's* test flight. A few thousand people showed up on the rainy holiday at an abandoned racetrack near Curtiss's workshop in Hammondsport. "The little half-mile track, partly overgrown with grass, had no grandstand, while a potato patch and vineyard crowded it uncomfortably close on both sides," Fairchild described the setting in his autobiography.

Spectators grew fevered with anticipation as they watched Curtiss make his first attempt. He succeeded in getting into the air, but he failed to fly far enough to win the prize. "The first flight had raised excitement to boiling point," Marian Fairchild reported to her father, who had stayed in Nova Scotia. "I don't think any of us quite knew what we were doing. One lady was so absorbed as not to hear a coming train and was struck by the engine and had two ribs broken."[6]

After careful preparation, Curtiss made his second attempt. "Suddenly we heard the roar of the propeller, saw the dust cloud which it raised behind it and then, against the pale gray of the evening sky, there came toward us this strange, white, flying apparition, with the long slender form of Curtiss out in front and a whirling thing behind him," David Fairchild recalled.[7]

Marian Fairchild was exhilarated when she described Curtiss' triumph to her father. "In spite of all I had read and heard, and all the photographs I had seen, the actual sight of a man flying past me through the air was thrilling to a degree that I can't express," she wrote. "We all lost our heads and David shouted, and I cried, and everyone cheered and clapped, and engines tooted."

Tom Selfridge was one of the excited bystanders who watched Curtiss win the trophy. At the end of the trial, he chatted with another man who urged him to beware of the risks of the new contraption. "You must be careful, Selfridge," he said, "or we will need a bed for you in the hospital of which I am a trustee." Selfridge assured the unidentified bystander that he would be. "Oh, I am careful, all right," he said.[8]

Years later, Fairchild remembered how amazed he was when he first saw Curtiss fly. "That brief afternoon in Hammondsport had changed my vision of the world as it was to be," he wrote in his autobiography. "There was no longer the shadow of a doubt in my mind that the sky would be full of aeroplanes and that the time would come when people would travel through the air faster and more safely than they did then on the surface of the earth."

The speed of air travel would eventually change plant exploring, too. One unromantic aspect of the work was the meticulous cleaning of seeds and packing of cuttings necessary to protect material during the weeks and months needed to travel to America by ship. Fruit seeds, for example, had to be completely clean of pulp and dried slowly in the shade to keep from spoiling in transit and being moldy on arrival. Many years later, Fairchild exulted when plants he found in the South Seas traveled on the Pan Am Clipper from Manila to Florida in only nine days.

*   *   *

By 1908 the Wright Brothers had abandoned their practice of flying only in private. Two months after the *June Bug*'s triumph, they agreed to compete against Bell's team for a contract with the U.S. War Department. Trials of each machine were scheduled at Fort Myer, Virginia, near the Arlington National Cemetery, in September 1908. To win the contract the plane would have to accommodate two people and stay in the air at least one hour. Wilbur Wright was busy in France at the time, so Selfridge volunteered to be Orville Wright's passenger on the test flight.

Wright's turn was originally set for Monday, September 14, but high winds delayed the test for three days. When the trial was finally set for Thursday, September 17, the change pleased Selfridge. He had to report to Saint Joseph, Missouri, the next day to fly an Army dirigible, so he would have missed his chance to fly with Wright if the test had been delayed any longer.

FIGURE 10. Thomas Selfridge and Orville Wright seated in Wright's *Flyer* on September 17, 1908, at Fort Myer, Virginia, shortly before the plane crashed and killed Selfridge. From the U.S. Air Force Historical Research Agency's archives branch.

Barbour Lathrop, who was a friend of Selfridge's father from San Francisco, was in Washington that day and arranged to have dinner at the Willard Hotel with Selfridge and David and Marian Fairchild after the trial. The Fairchilds stayed home in Maryland during the day as Lathrop accompanied Selfridge to the testing grounds. About 1,500 people traveled to the big event, filling the streetcars from downtown Washington to Fort Myer.

Lathrop watched as the Wright-Selfridge plane took off. The test began well as the plane made three circles above the excited crowd, which cheered them on. One witness said, "I was standing in the part of the field nearest the homes of the army officers and evidently a number of the friends of Lieutenant Selfridge were in the group near me, for as the airship passed over us, Lieutenant Selfridge took off his cap and waved it at a number of the ladies and men in the party. . . . He was smiling and laughing."[9]

Suddenly, on the fourth lap, something went wrong. A piece of a propeller fell off the plane, and the machine's nose immediately tipped toward the ground, dragging the plane down in a short, sharp dive from about seventy-five feet. Wright heard Selfridge say softly, "Oh! Oh!" before the plane crashed and the two men hit the ground. "They were thrown in the air and then fell forward on their faces," a witness said.

Wright broke his leg and three ribs in the crash, injuries that caused permanent nerve damage. Selfridge's skull was fractured. He never regained consciousness and was pronounced dead later that night at a downtown hospital. At twenty-six, Tom Selfridge had become the first person in history to die in an airplane crash.

David and Marian Fairchild were dressing to meet Selfridge and Lathrop for dinner when they heard about the accident. They went to the hotel, but Lathrop arrived at the Willard without the triumphant aviator they had expected. They soon got word of his death. "We sat down to dinner in very low spirits indeed," Fairchild wrote. It was Fairchild's sad duty to send a telegram to Alec Bell notifying him that his promising young assistant had died.

Reporters tracked down Lathrop to interview him about the accident. The old journalist gave them a good firsthand account and tried to be positive. "He had his heart set on flying with Mr. Wright at Fort Myer," Lathrop said. "He realized his desire and paid the price with his life. I fancy few men have spent their last hour in keener enjoyment."

Regardless of Lathrop's upbeat remarks, Selfridge's tragic death devastated him. "Mr. Lathrop's nerves were thoroughly shaken, and for years afterwards he never cared to see a flight," Fairchild recalled later. "He called the airplane 'a thing of rags and tatters.'" And, throughout his life, he never once traveled by airplane.[10]

A Beautiful Job

After David Fairchild returned to Washington from his odyssey with Barbour Lathrop and fell in love with Marian Bell, his down-to-earth work began in earnest. He resumed the job he had abandoned in 1898: managing foreign seed and plant introductions for the U.S. government.

By 1904 the responsibilities were bigger than ever. The agency had three staff explorers in the field: Ernst Bessey was in Russia and Turkistan looking for alfalfas, Archibald Shamel was in Cuba collecting tobacco, and Thomas Kearney was hunting for new varieties of dates and olives along the Mediterranean coast. By then the federal government was prepared to test the explorers' material. It had set up a plant introduction garden in Chico, California, near Sacramento, where gardeners conducted trials on seeds and cuttings before distributing the successful ones to likely commercial growers.

After the program was organized, however, one major task remained: exploring the mysterious center of China. Although the vast terrain was not completely unknown, the job remained an irresistible challenge for lovers of plants and exotic places. Finding someone to do the job was not easy.

Since the middle of the nineteenth century, a handful of western botanists had visited a few corners of the country. The most experienced was Augustine Henry, the Scottish doctor who had urged Fairchild to send his own man rather than depend on missionaries.

No American had ever hunted for plants in China's interior. British companies and botanical organizations, eager to track down new plants for their customers, had dispatched a few experts to the region. In 1842, after the Opium Wars settled down, the Royal Horticultural Society had hired a young Scottish botanist named Robert Fortune as its explorer. He returned several times—twice for the East India Company and once briefly for the U.S. Department of Agriculture to look for tea seeds in 1858—but Chinese officials never allowed him to go beyond port cities and cultivated gardens. Although he sometimes disguised himself as a Chinese man—his outfit included a long braid down his back—Fortune couldn't roam freely through the countryside exploring for the wild, unfamiliar plants that would be the most promising introductions.

After Fortune, several French missionaries who were serious botanists managed to capture hurried glimpses of the interior. An important breakthrough came in 1869 when Father Jean Armand David discovered a tree covered with pure white flowers that was unknown in Europe. Botanists quickly spread the news that the most beautiful tree in the world had been spotted in China. They named it the Dove tree (*Davidia involucrata*) because its gently fluttering white flowers resembled birds at rest.

Twenty years after Father David's discovery, Augustine Henry found a second Dove tree in a different province, more than doubling botanists' enormous interest. Ten years after that, James Veitch and Sons, a big British nursery company, hired Ernest Henry Wilson, a twenty-three-year-old gardener at the Royal Gardens at Kew, to collect specimens of this almost-legendary tree. Wilson spent three years exploring China and, in 1902, triumphantly returned to England with seeds of the Dove tree as well as other dazzling ornamental plants. "I am convinced that *Davidia involucrata* is the most interesting and most beautiful of all trees which grow in the north temperate regions," Wilson wrote later.[1] (Gardeners still consider the tree to be remarkable, although it grows so slowly in North America that one nursery, Avant Gardens in Dartmouth, Massachusetts, recommended in 2013 that customers plant them for their grandchildren.)

By 1904, when Fairchild returned to supervising foreign plant introduction, he was determined to get someone inside China. The ideal candidate needed sophisticated botanical knowledge and lots of stamina. Fairchild had a problem, however; none of the qualified people wanted the job.

Fairchild asked Augustine Henry to return to China for the U.S. government. He said no. Fairchild next asked Ernest Wilson, but he refused, too. Fairchild was so eager to investigate the Chinese interior that he considered going himself, but that plan collapsed about the time he became engaged to Marian Bell.

While these experienced China hands were rejecting Fairchild's job offer, Frank Nicholas Meyer, a young Dutch immigrant, was working as a laborer at the Missouri Botanical Garden in St. Louis. After arriving in America from Europe four years earlier, Meyer had worked briefly for the agriculture department as a gardener in Washington and at an experiment station in California. Too restless to stay in any job for long, Meyer kept quitting his job and moving on. Two facts were evident wherever he went, however: he loved plants, and he loved to walk.

Born Frans Meijer, he had grown up in Houthaven, the port of Amsterdam, in a modest house with a small garden. Even as a young man, he was restless and sensitive, and he embraced reincarnation as his fundamental religious belief. "It is strange that I have to wander so much," Meyer wrote after he moved to America. "I must have lived very strange in my previous life."[2]

When he was a young man Meyer left Holland on foot and hiked across the Alps to Italy—in the middle of a blizzard—to see what orange trees looked like. When he arrived in Italy, the first person he met asked him where he had come from.

"From over the mountains," Meyer answered.

"Impossible," the Italian said. "There are no roads."[3]

Roads didn't matter to Meyer. In 1903, during a break between jobs in America, he spent fourteen days walking 260 miles through the Mexican countryside from San Blas to Guadalajara. The first thing he did after he arrived in the nation's capital was climb all 898 steps of the Washington Monument.

Meyer's botanical training started early. After quitting school at the customary age of fourteen, he took a job at the Amsterdam Botanical Garden where the director was Hugo De Vries, a botanist of worldwide reputation. Garden staff members assigned Meyer menial chores, but De Vries quickly noticed his talents and made him his personal assistant. De Vries taught him English and French and paid for his botany studies at the University of Groningen. Meyer stayed with De Vries for nine years.

Yet he remained restless, always yearning to find a better place. On the day Meyer turned twenty-three, wanderlust overcame his affection for his mentor; he quit his job and began a nomadic quest for personal fulfillment. He lived for a year in a utopian community near Amsterdam called Walden. That experience disappointed him because the other utopians lacked the spirit of true brotherhood. Meyer complained that the nonintellectuals wouldn't share the cost of kerosene with the intellectuals who wanted to read after dark. "The serpent of selfishness was there," Meyer remarked.[4]

After that disappointment, he gave up on Holland and decided to move to the United States, a land that had beckoned to him since he had read James Fenimore Cooper's books as a boy. As he prepared to leave Holland and begin life in a new country, Meyer was hopeful about his future. "I live now in expectation of what will come," he told friends.[5]

Meyer's ship, the SS *Philadelphia*, carried him from Southampton to New York. He arrived on October 19, 1901, and immediately boarded a train to Washington. A letter of introduction from De Vries got him in to see Erwin Smith, the plant pathologist at the agriculture department who had spotted Swingle's intelligence several years earlier. Smith immediately hired Meyer as a gardener at the government greenhouses on the Mall, near the present site of the National Gallery of Art.

Meyer's hard work and devotion to plants impressed his co-workers. He stayed in the job for less than a year—one summer in hot, muggy Washington, D.C., was enough for this northern European—and then he went west. He worked for more than three years at various gardens and commercial nurseries in the United States and Mexico. He was still restless, always walking and always fascinated by plants. "I know quite a few people frown upon my roaming around," he wrote to a friend after he quit one job to explore Mexico, "but I personally have learned more about the true nature of plants during this two-months' trip than all the books and hothouses could have brought me in ten years."[6]

His government jobs led to several lifelong friendships, most significantly one with Adrian J. Pieters, another botanist of Dutch background who worked with Fairchild in Washington. This bond was key to Meyer's success; in the spring of 1905, Pieters convinced Fairchild that Meyer was the man he was searching for.

FIGURE 11. David Fairchild and Frank Meyer seated at Fairchild's desk in Washington during a break in Meyer's expeditions to Asia. From the Frank N. Meyer collection, National Agricultural Library. By permission of Special Collections, National Agricultural Library.

Meyer was working in the botanical garden in St. Louis when he received a telegram offering him a new job: plant explorer in China for the U.S. government at an annual salary of one thousand dollars plus expenses. On March 11, 1905, Meyer wired back: "Great thanks for your offer. Accept it." Fairchild summoned him to Washington in July for a meeting that was important for both men.

David Fairchild was no longer the gawky young scientist Barbour Lathrop had discovered fumbling at his desk in Naples. He was thirty-six and associated by marriage with one of America's most revered inventors. He was directing a new government agency engaged in groundbreaking work. He had an impressive job title: "Agricultural Explorer in Charge of Seed and Plant Introduction." And he was on the verge of making his first significant decision, selecting a full-time explorer for China.

Meyer, thirty, was ready to stop wandering aimlessly and determined to find his mission in life. He had worked with plants for sixteen years and had decided that finding better food for Americans suited his idealistic, independent spirit. He was strong and fit, with melancholy blue eyes and a full dark-brown beard.

When the two men first met in Fairchild's office, the weather was uncomfortably hot. "Meyer was one of those full-blooded men who had spent his life out of doors and perspired freely," Fairchild recalled. "He cared nothing about his dress. Somewhere he had picked up a striped shirt, and when he came to see me it was wringing wet and the stripes had run. But he sat on the edge of the chair with an eagerness and quick intelligence that won me in an instant." Meyer told Fairchild that when he was working at a government experiment station in California, his boss refused to let him use mulch to protect several bamboo plants that Fairchild and Lathrop had sent from Japan, and the plants had died. "As Meyer told me about it, his eyes filled with tears," Fairchild said.[7]

He was impressed not only by Meyer's deep affection for plants but also by his passion for walking, an essential requirement for exploring a massive, rough country without decent roads. Meyer preferred walking to any other form of transportation. "I can see him now as he held up a new pair of walking boots he had bought and in which he took more pride than many a young man would in his first automobile," Fairchild remembered later.[8]

Fairchild's brief visits to Canton had been exciting but comfortable, as travel always was with Barbour Lathrop. Accompanied by a well-informed Chinese guide, Fairchild had traveled by sedan chair, a wicker seat suspended on poles carried on men's shoulders. Because Fairchild's own expeditions were financed by his mentor's large fortune, he had no firsthand knowledge of the rough conditions in remote areas of China. Even Augustine Henry, Ernest Wilson, and the other explorers had enjoyed generous budgets and the backing of enterprises with solid experience in China.

Fairchild apparently assumed that Meyer would operate under conditions similar to his own. Twelve years earlier, when Fairchild was a passenger on the *Fulda* on his first trip to Europe, he had cowered behind a pillar in the ship's salon because he didn't have the formal evening wear required in the dining room. Fairchild expected that Meyer would need formal clothes in China, too, so he presented him with a tuxedo as a farewell gift. Meyer packed it carefully on the bottom of his steamer trunk.

Fairchild gave Meyer little time to prepare for his first expedition. He sent him on a quick ten-day trip to the New York Botanical Garden and Boston's Arnold Arboretum to familiarize himself with plants and trees that had already been introduced from China. Meyer briefly visited the experiment station in Chico to meet the gardeners who would cultivate the seeds and cuttings he sent them. Less than one month after meeting Fairchild in Washington, Frank Meyer was onboard the SS *Coptic*, steaming out of San Francisco Bay and bound for Shanghai and his new life.

*   *   *

Although the plant exploring opportunities were extraordinary, the assignment was an enormous challenge. While China's climate and soil are similar to those of North America—especially in the northern reaches—the territory was a vast landscape filled with unfamiliar trees, plants, and flowers.

The rewards of exploring could be great. Because the Chinese had farmed for thousands of years, they had developed insights that might help American farmers desperate for new crops and better techniques. Giant peaches, disease-resistant pears, and mighty chestnut and walnut trees were a few items rumored to grow in the mountains and valleys deep in China. Fairchild assigned Meyer to locate plants at the right time of year for gathering seeds and cuttings and somehow transport them—alive—halfway around the world to Washington.

At the start of the twentieth century, China was a vast territory—not a nation—that covered more than 3.5 million square miles. The size of its population was unknown. It had no single language or stable currency. When Meyer arrived, the Manchu dynasty was struggling to maintain the absolute power it had held for more than 250 years. As the government's control weakened, anarchy spread through the country. Sadistic bandits terrorized large sections of China, including several areas with the most promising plants. Poverty and execrable health conditions were widespread.

Yet when Meyer left America in the summer of 1905, he was optimistic about the adventures that lay before him. He wrote to his family on August 2, 1905: "I can hardly believe I got such a beautiful job."

*   *   *

Frank Meyer's first task was to scour north China (a vague term that covered Mongolia, Manchuria, eastern Siberia, northern Korea, and the area around Peking) for hardy fruits that could grow on the Northern Plains. As a botanist without a reference library, he depended on his brain and senses to distinguish between known and unknown varieties. Meyer had memorized botany encyclopedias rather than lug heavy books, and he carried thousands of plant images in his mind. He constantly compared them with the live foliage he saw as he walked along narrow, unpaved roads. "It was characteristic of Meyer to quickly break off a fragment of a plant and chew it or smell it furiously in order to bring to bear upon it all the powers of his remarkable memory," Fairchild wrote.[9]

Meyer worked under trying conditions. His expenses were limited to $2,500 a year; his salary was $1,000. He did not speak any Chinese language. He traveled without western companions. No one, not even Augustine Henry, had ever attempted to do what he was going to do.

At first, Meyer stayed close to Peking. He drank tea with friends of Fairchild and made contacts with western missionaries and U.S. diplomats. He saw a few legendary sights: his brief trip to the border of Inner Mongolia included a glimpse of the Great Wall's eastern extremity. Another quick tour southwest of Peking took him to the first Ming emperor's 500-year-old tomb.

Meyer enjoyed the novelty of displaying his common western possessions. A flashlight with its on-off switch intrigued Chinese villagers; others were mystified by the smell and taste of ground black pepper. He also noticed filthy conditions and became wary of people's suspicion of him as a "western devil." Within a month of arriving, he was forced to spend the night in a dirty inn in Mongolia and struggled to fall asleep in the typical bed, a four-foot-high pile of bricks infested with fleas. Meyer often attracted a large audience when he undressed and washed in a public bathhouse, yet he accepted these conditions with patience and good nature. "I had to dry and dress in great show just to let them see how we do it," he reported. "Happily no buttons came off, so I hope I left a good impression."[10]

He was less good-natured about Washington's bureaucratic demands. One time, after Meyer paid seven Chinese boys 20 cents each to collect seeds for him, he was furious when the agriculture department refused to reimburse the $1.40 because the children, who were illiterate, hadn't signed the necessary vouchers.[11]

* * *

In February 1906, six months into Meyer's first expedition, he faced a more serious situation. Early in the month he heard a rumor that all foreigners in China would be massacred on February 25. It didn't happen, but three days later thugs in Hankow mugged him while he was returning to his hotel after dinner. Meyer's attackers punched him in the back while bystanders hooted and sneered instead of rescuing him. Still, he handled the incident gracefully. "I took this howling for an ovation, and took my hat off and bowed in all directions and smiled like a president on his inauguration trip," Meyer told friends. "These little incidents add zest to travel."[12]

Meyer did manage to find a few chances to relax. On a steamer traveling down the Yangtze River from Hankow to Shanghai in February 1906, he met other westerners and swapped adventure stories with them as they drank wine and beer together. Meyer boasted about his visits to inland areas of China rarely seen by foreigners. On this boat trip he was affected deeply by his first view of the mighty, magnificent Yangtze, calling it a splendid body of water.

When he arrived in Shanghai, a western stronghold that looked like Philadelphia, the U.S. government paid for him to stay in a comfortable room with a soft bed. Meyer bragged to his family that he was living like "a millionaire at the Astor Hotel."[13]

* * *

That spring Meyer set off for Manchuria, his first long trip inside Asia. It was a remote but promising destination because Manchuria's growing conditions were similar to those of the northern United States, the section of the country that Secretary Wilson wanted most to help.

Problems plagued the trip from the beginning, however. Officials wouldn't let Meyer travel freely because Russian and Japanese soldiers were still skirmishing in the region, a bitter after-effect of the Russo-Japanese War that had ended only seven months earlier. Notorious outlaws called the Hun-hutzes (Red Beards) also menaced the area.

Despite these obstacles, Meyer, confidant he would be safe, was determined to make the trip. He knew he could be physically intimidating, especially when he wore a heavy sheepskin coat, big boots, and a bearskin hat to survive temperatures that dropped to 30 degrees below zero Fahrenheit. With a revolver and a Bowie knife in his belt, Meyer was prepared to defend himself. He relished the adventure. "I talked with fellows who showed me scars of wounds they had received in combat with these chaps," he wrote Fairchild as he left in early summer 1906, "so I have some interesting trips ahead of me."[14]

The local laborers he hired were not as sanguine. After a few weeks of travel, his few Chinese employees were terrified by the cries of tigers in the night and by reports that robbers along their trail had roasted a Chinese man alive. They quit.

These setbacks forced Meyer to cut the length of his expedition in half. He spent only three months in Manchuria, including side trips to northern Korea and Siberia. It was still a rough expedition: he covered 1,800 miles from Liaoyang to Vladivostock almost entirely on foot, averaging twenty miles a day for ninety days. He wore out three pairs of boots in three months. On the way he saw beautiful peonies growing wild and collected many specimens of useful plants, including one that eventually became enormously important to America: the soybean.

Before Meyer went to China, fewer than a dozen soybean varieties existed in America; most experts credit Benjamin Franklin with introducing an early one from France. Niels Hansen sent another type from Siberia in March 1898, but few Americans were interested in the plant's potential. But Meyer, recognizing that it was a mainstay of the Chinese diet, sent samples to Fairchild: he collected seeds, whole plants, even beans prepared as tofu, which he called cheese. During his travels Meyer shipped more than one hundred varieties—including ones that launched America's vast soybean oil industry.

During this first visit to Siberia in 1906, strangers attacked Meyer again, nine months after the first incident in Hankow. As he was returning to his

FIGURE 12. Frank Meyer dressed to stay warm and to intimidate attackers in China. From the Frank N. Meyer collection, National Agricultural Library. By permission of Special Collections, National Agricultural Library.

inn after dinner on a frigid November night, three men jumped him in the dark. "I was suddenly grabbed from behind by a pair of powerful horny hands, which tightened upon my windpipe and so brought me to the verge of strangulation," he told the *Washington Post* afterwards.[15] Meyer fought back. He wasn't carrying his revolver that night, but he pulled out his knife and stabbed one attacker in the belly. The three muggers fled. Meyer was not injured, but the next morning police found the body of another man near the site of the attack, apparently a victim of the same thugs.

About this time, Meyer told De Vries that he had wanted to walk across Manchuria to Harbin, but the trip would have been too dangerous, so he took a train. Tigers, panthers, bears, and wolves lurked nearby, but Meyer said he was more afraid of humans than wild animals. He usually carried with him seven sacks filled with currency: Mexican, Hong Kong, and Hupeh dollars, silver and copper coins, and lump silver. (The last he spent by slicing off the amount he needed with a file.) Meyer was a tempting target. "The danger of robbers is greater," he wrote. "I carry a revolver and knife or keep a weapon under my pillow."[16]

Extreme weather was another problem. In January 1907 he told his sister that the temperature in Harbin was 25 below zero Fahrenheit. He wore two pairs of trousers, an undercoat, a sheepskin coat, a bearskin hat, a scarf, earmuffs, and a sheepskin nose cover. Yet if he sat still for a half hour he would nearly freeze to death.[17]

* * *

By this point Frank Meyer wasn't the only man exploring China for American institutions. At the beginning of January 1907 Ernest Wilson returned to the country as an employee of Harvard University's Arnold Arboretum to hunt for trees, flowers, and dried specimens. To pool their resources, Fairchild and Charles Sprague Sargent, the Arboretum's director, coordinated their explorers' work. Sargent promised Fairchild that Wilson would look for hardy food plants for the government if Meyer would collect ornamentals and dried samples for the arboretum.

At this time, a new translation of Marco Polo's account of traveling through China in the thirteenth century was published in the United States, reviving interest in his adventures in Asia. Polo described a lush

FIGURE 13. Ernest H. Wilson, right, with Walter Zappey and two dogs in China in 1908. Photo from the Arnold Arboretum Horticultural Library of Harvard University, © President and Fellows of Harvard College. Arnold Arboretum Archives.

area of Shansi province southwest of Peking that was filled with grapes that the Chinese made into wine. He also wrote about beautiful mulberry trees in Shensi, an adjacent province.

Sargent, working in his comfortable office in Jamaica Plain, Massachusetts, persuaded Fairchild that Meyer should retrace Marco Polo's route and explore Wutaishan, a mountainous district of Shansi province. Meyer followed orders, but when he arrived in Wutaishan in April 1907, he discovered that the Chinese had so completely deforested the once-verdant region that it was "as barren as the plains of Nebraska."[18]

Meyer photographed the desolate valley and collected a few cuttings from temple gardens, but he was angry because the side trip wasted his time and interrupted the collecting he wanted to do in Manchuria. He was also furious after he learned that Ernest Wilson worked under conditions far better than his own. Despite his romantic job title and the thrills of international plant exploring, Meyer was actually a low-paid government functionary who had to account for every penny he spent. Wilson was not. His budget for two years, paid by wealthy patrons of Harvard's Arnold Arboretum, was thirteen thousand dollars, more than twice Meyer's allotment. Meyer's mission was to help industrious farmers make a living in the rough conditions of the Great Plains. Wilson's was to find interesting trees, ornamental plants, and specimens for scholars to peruse in the quiet reading room of Harvard's herbarium.

Wilson could afford to establish a comfortable base in Ichang, an especially rich and fertile region of China. His habit was to reconnoiter in the vicinity for two or three weeks, often with an entourage of twenty-five workers. Because Wilson believed it was important for western visitors to save face in China, he always traveled with a sedan chair to advertise his superior rank.

"A sedan chair is an outward and visible sign of respectability," Wilson wrote in an account of his adventures. "It is the recognized medium of travel and, quite apart from its real use, it is a necessity since its presence ensures respect. In the out-of-the-way parts of China, even though it is carried piecemeal, a chair is of greater service and value to the traveler than a passport."[19] (It wasn't always a safe way to travel, however; Wilson was riding in a sedan chair in Chengu in 1910 when he got caught in a rock avalanche and badly injured his leg.)

Meyer, never an elitist, scorned this habit. He refused to allow workers to carry him, although the decision sometimes cost him the respect of Chinese people he met on the road.

Because Wilson had explored the area around Ichang for Veitch a few years earlier, he had a team of trained local men in place there, including collectors and guides who understood him and his needs. Like Meyer, of course, Wilson sometimes slept in tents or on flea-infested brick beds in roadside inns. But his overall situation was far more comfortable. For the first two years of Wilson's expedition, Walter Reeves Zappey, an American scientist from the Museum of Comparative Zoology who was hunting for birds in China, traveled with him as an agreeable companion. During his first assignment from Sargent, Wilson set up base in a two-room houseboat that he called the *Harvard*.

The more Meyer learned about Wilson's circumstances, the more resentful he became. Soon he was making demands. "If I am going to stay in this line of work in the future," Meyer angrily wrote to Fairchild on February 9, 1908, "I demand to be treated as, for instance, Wilson has been treated, being sent to regions which are well-known to be rich in vegetation and not to be expected to reap a bountiful harvest in the desert."

Fairchild immediately reassured Meyer that his work was valuable to America. "You have the ideal spirit of an explorer," he told him. "I do not believe that any other explorer has ever been as successful as you have been in landing in living condition in this country such a large number of introductions and valuable scions and plants."[20] Fairchild's compliments soothed Meyer, at least temporarily, so he continued to collect everything he could. He soon made a discovery of his lifetime.

On March 31, 1908, as he was heading to Peking toward the end of his first expedition, Meyer stopped briefly in the small village of Fengtai. In a doorway he noticed something he had never seen. It was a small tree bearing about a dozen unusual fruits that looked like a cross between a lemon and an orange. Villagers told him that the strange plant was valuable; rich Chinese paid as much as ten dollars for each tree because it produced fruit all year. "The idea is to have as many fruits as possible on the smallest possible plant," Meyer explained later.[21]

He sliced a thin branch off the tree with his Bowie knife and packed it carefully in damp moss. Meyer delivered it two months later to Fairchild. He gave the cutting an unexciting label—"Plant Introduction No. 23028"—and sent it to the department's garden in Chico, California, to see if it would grow and, what was more important, produce fruit in America. The experiment lasted seven years, but eventually Fairchild was able to report that the cutting was a success. "Meyer's dwarf lemon from Peking was producing a high yield," he said. "It had begun to attract attention as a possible commercial lemon, even though its fruit flesh had an orange tint."[22]

Although it took scientists several more years to figure out how to protect the Chinese import from domestic plant pests, California farmers eventually cultivated hundreds of acres of the fruit, mostly for its generous juice. Then creative chefs, especially Alice Waters of Chez Panisse in Berkeley, discovered its unusually sweet taste and invented dishes that showcased its flavor. One hundred years after its discovery, the Meyer lemon has become a culinary star in America, an accomplishment that fulfilled the promise he made to his family in 1906. "I will be famous," Meyer vowed, "just wait a century or two."

Six weeks after he spotted the lemon, Meyer boarded a ship in Shanghai for San Francisco. He carried twenty tons of trees, cuttings, seeds, and dried herbarium material as well as, almost as an afterthought, two rare monkeys for the National Zoo. "They cause me as much trouble as babies," Meyer complained when he arrived in California in June 1908.

He also returned with a few personal souvenirs that shocked his government colleagues: his dark hair had grown to his shoulders, and his long fingernails curled backwards in the style of a Mandarin.

*   *   *

Fairchild wasn't the only American who was pleased about Meyer's work. Later in the summer of 1908, Henry L. Hicks, a commercial nurseryman from Westbury, Long Island, took Meyer to meet President Theodore Roosevelt at Sagamore Hill, his home in nearby Oyster Bay. Roosevelt, who was battling with Congress over the need for tough conservation laws, wanted a firsthand account of the devastation of Wutaishan.

The burly plant explorer, seated in a leather armchair in a large room decorated with moose heads and bearskin rugs, described deforestation in China to the president of the United States. "The Chinese peasants have no regard for the wild vegetation and they cut down and grub out every wild wood plant in their perpetual search for fuel," Meyer explained later.[23] After their talk, Roosevelt praised Meyer and his employer. "Of the good work that was being done in the country by different institutions and men, that of the Department of Agriculture ranks in my opinion among the foremost," the president told Meyer, who reported the compliment to Fairchild on September 12, 1908.

Four months later, in the leaflet he sent to Congress as his annual State of the Union message, Roosevelt quoted Meyer by name and included his photographs to illustrate the price America could pay if the nation didn't protect its trees. Meyer's pictures, Roosevelt told lawmakers, "show in vivid fashion the appalling desolation, taking the shape of barren mountains and gravel and sand-covered plains, which immediately follows and depends upon the deforestation of the mountains."

Charles Sargent, who had sent Meyer to Wutaishan to record lush foliage, must have been annoyed by the attention paid to Meyer's work. Sargent was a proud, difficult man who demanded much of his associates. "I voted him autocrat of autocrats," Ernest Wilson said after their first meeting.[24] Fairchild found him irritating, too, largely because Sargent belittled Fairchild's work and once said dismissively that government botanists didn't care about any plant they couldn't eat.[25]

A few months after the president's speech, Sargent sent Fairchild a crotchety note announcing that Meyer's collecting had displeased him so much that he was charging the U.S. government $5 (the equivalent now of about $120) for each plant Wilson had collected for Fairchild. Fairchild sent him the money, but later he got back at Sargent. He wrote in his autobiography that he hadn't been satisfied with Wilson's work, either, adding that Ernest Wilson, the internationally celebrated plant explorer, once mailed him a few grains of barley erroneously labeled "wheat," an unforgivable mistake for a serious botanist.

* * *

During his travels Meyer wrote hundreds of letters from the field that his colleagues in Washington read avidly. Because Fairchild was eager to promote his work to growers and botanists around the world, on May 1, 1908, he started publishing a regular government bulletin he called *Plant Immigrants,* again using the phrase Bell had suggested. Fairchild featured photographs and passages from Meyer's letters in each issue. "I traveled with him, in spirit if not in body, through the farms, gardens, forests, and deserts of Asia," Fairchild wrote later.[26]

To celebrate Meyer's triumphant return to Washington in the summer of 1908, Fairchild invited the office staff to a barbecue at his home in Kensington, Maryland. On the clear, moonlit night, Fairchild built a huge bonfire. Food was roasted. Someone spiked the cider. Meyer, who had often been lonely in China, relaxed among his new American friends and associates. "As he lay stretched out on the ground before the dying embers of the bonfire, he sang us his favorite Dutch songs and gave us an account of the uncomfortable brick beds, the vegetable gardens, and the poor agricultural class of China," Fairchild recalled.[27]

Everyone in the office, both in the field and in Washington, had contributed to the success of the department's first China expedition. The bonfire party was a happy memory for Fairchild and Meyer.

proof

Easy Money

Editors at the *Los Angeles Times* considered Frank Meyer's work to be newsworthy, even noble. During his years as an explorer, the newspaper published half a dozen stories about his adventures in Asia, as well as accounts of Ernest Wilson's travels. One avid reader of the stories was Frederick Wilson Popenoe, a teenager living in Altadena, California. "I began to feel that plant hunting was just about the most romantic occupation imaginable," Popenoe wrote later. "Not only did a chap get to travel in out-of-the-way corners of the world, but he stood a good chance of bringing home some new fruit or food plant, which would add materially to his country's wealth and happiness."[1] Popenoe was so fascinated by plant exploring that it took him only a few years to join Fairchild's team and become his protégé.

Like many of his colleagues, Popenoe came from the rich farming lands of Kansas. His grandfather Willis Parkison Popenoe, the man who had said that grasshoppers seemed to cover the face of the earth in the 1870s, made his living growing apples and other fruit in a nursery near Topeka. His uncle taught entomology at Kansas State. Fairchild first met Wilson Popenoe when he was a thirteen-year-old student poking around his uncle's college laboratory.

Wilson Popenoe's father, Frederick Oliver Popenoe, was born in Illinois during the Civil War. At twenty-four he married Marion Bowman, the wealthy daughter of a Topeka businessman. In the beginning, resisting the forces that surrounded him, Popenoe tried to make a living doing anything but farming. For a while he was the Kansas governor's official stenographer; later he worked for a loan company owned by his father-in-law. He and his wife had three sons.

By the end of the nineteenth century, Fred Popenoe had acquired all the trappings of a solid midwestern capitalist. He lived with his family in a large house on the outskirts of Topeka. He was president of the Accounting Trust Company; by August 1899 he was also owned a local newspaper, the *Topeka Daily Capital*. Before long, however, Popenoe's conventional side gave way to his unconventional one.

When Fred Popenoe took over the newspaper, one of Topeka's most celebrated residents was the Reverend Charles Sheldon, author of a wildly popular novel, *In His Steps*, published in 1896. The book told the story of a church congregation whose members vowed to live their lives for one year as they thought Jesus Christ would have. The question that guided their behavior was, "What would Jesus do?" One character was a newspaper editor who always asked himself whether Jesus Christ would approve of an article before he published it.

Early in 1900 Popenoe met Sheldon at a local parade and, in a daring bid to boost circulation, offered him the opportunity to reenact the fictional editor's vow. Sheldon accepted and became editor-in-chief of Topeka's daily paper for one week beginning March 13, 1900.

Heeding his characters' slogan, Sheldon decided that Jesus would drastically change the *Topeka Daily Capital*. Sheldon banned liquor ads and stories about prizefights and other vices. He made reporters clear their copy with the people they wrote about before the stories went to press. He refused to run articles about the stock market because he considered investing to be gambling. Finally, he included only articles from or about the Bible in the Sunday edition and distributed it on Saturday so no one had to work on the Sabbath. "I think it may safely be said that after recovering from the shock caused by getting a paper without any news in it, many of the subscribers read for the first time, perhaps, the whole of the Sermon on the Mount," Sheldon wrote later, "and it may have been news to some of them."[2]

The gimmick worked, temporarily. In normal weeks the *Daily Capital* had an average daily circulation of about 1,800, but as newspapers around the world reprinted the special issues, it briefly soared to more than 360,000 under Sheldon's editorship. Popenoe wanted to keep running the paper Sheldon's way, but after the first week his staff revolted and advertisers fled. The publicity stunt eventually cost him the newspaper and almost everything else he owned, according to an unpublished history of the Popenoe family.

He managed to hold onto one asset: an interest in a Costa Rican gold mine he had bought in 1896. Popenoe, still a promoter, insisted that the investment would pay off. "The mineral wealth of Costa Rica is extensive and so far untouched," he said in a local interview published on April 4, 1897. "A fortune can be made in that country by anyone who has some capital to begin with, plenty of judgment and perseverance to continue, and a willingness to put up with untold inconveniences. No others need apply."

After he lost his home and his business in Topeka, Popenoe moved his wife and sons to San Jose, Costa Rica, to work full time running the mine. The three boys loved living in the tropics. They played in the jungle, ate their first avocados and other tropical fruit, and learned a little Spanish.

Fred Popenoe's experience was not as idyllic. In his desperate push to strike gold, his fortunes deteriorated further. After several failed investments, everything was gone: his own money, his wife's dowry, and finally at least $15,000 he had borrowed from friends and bankers in Topeka. By 1906 the mine had gone bust, and the family had no home and no money.

Fred Popenoe returned to America to find work; he was hunting for a job in San Francisco on April 18 when he narrowly escaped death as his hotel collapsed in the great earthquake. Soon afterwards he moved his family to southern California and never returned to Topeka, probably because he hadn't repaid the money he borrowed.

Wilson Popenoe said the tropics cost his father "not only his fortune, but his health."[3] His mother apparently took her husband's dramatic financial failures with grace and humor. After another of Fred Popenoe's get-rich-quick-schemes collapsed, she told one son, "Oh, well, all that money would probably have made you a snob."[4]

*   *   *

The Popenoe family started over in Altadena, a rural community south of Los Angeles. While his father began rebuilding his life, Wilson, who was about fourteen, set up a small garden behind the family home. He grew rose bushes and other plants and hawked them from a wheelbarrow to wealthy neighbors. On each sale, he made a profit of $1.50, "just about the easiest money, it seemed to me, that anyone could hope to earn," he bragged.[5]

Fred Popenoe had a similar idea, but on a larger scale. In 1907 he started a commercial nursery called the West India Gardens and found himself back on the path set by his own grandfather in Kansas: selling plants to make a living. This time Popenoe's gimmick was marketing exotic plants to local rich people. His first big investment was the avocado, which was then called the alligator pear. This unusual fruit was becoming a popular salad ingredient in southern California's fancy restaurants. In 1911 Popenoe sent an employee named Carl Schmidt to Mexico to find new stock for his nursery. Popenoe wasted no time promoting the plants that Schmidt discovered. "The avocado is the most valuable fruit grown," Popenoe asserted in a 1911 publicity booklet. As he had with his Costa Rican gold mine, he promised customers they would get rich doing business with him. "Within the next quarter century, the avocado will rank with the orange as a commercial fruit of southern California," he wrote.[6]

One especially tasty variety Schmidt found in Atlixco, Mexico, was initially known only as Number 15, but Wilson Popenoe changed its name in January 1913 after a freak frost hit Altadena. The cold killed all the avocado varieties in West India Gardens—except the tasty one from Atlixco. Popenoe named it Fuerte because it had been strong in the face of cold temperatures. It stayed that way, too; for many years the Fuerte was the most popular avocado grown in America. Nonetheless, West India Gardens lost $100,000—or well over $2 million today—on Popenoe's avocado scheme.

Despite this commercial failure, Fred Popenoe deserves much of the credit for introducing avocados to America. Years later Schmidt praised him for having the vision to send him to Mexico to collect the plants. "Popenoe was a nut, an imaginative, idealistic nut," Schmidt said. "In 1911 his current nutty idea was that California would be a good place to grow avocados—and that people would like them."[7]

* * *

The same year Popenoe sent Schmidt to Mexico, he had another big idea: the date palm. By then Walter Swingle's early confidence that the trees would flourish in the California desert had been realized; in 1912 the *Tacoma Times* newspaper said that the California Date Company in Heber, the firm Swingle had organized, had developed the finest plantation of imported dates in America. Farmers throughout the region decided that selling date palm offshoots was another excellent way to make money. A date palm boom—a phenomenon not unlike a gold rush—was under way. "Promoters of stock-selling schemes have rushed in, sending out mail sacks full of literature and making the most extravagant promises of profits to the investors for their shares," reported Silas Mason, a colleague of Fairchild.[8]

One person lured by these claims was Rebecca Lee Dorsey, a respected Los Angeles doctor who asked Fred Popenoe to find her enough offshoots to start a grove on property she owned near Indio, California, a desert town about 125 miles from Altadena.

After he agreed to supply the material, Popenoe directed his oldest son, Paul, who was on a walking and cycling tour of Europe, to visit North Africa to buy good varieties. Before Paul began his expedition, Popenoe arranged for him to meet Walter Swingle in Spain. Swingle, courteous and helpful but still unable to cash in on the date boom himself, told Paul where to go and what to look for. He also gave him a handy Arabic phrasebook.

Paul spent several weeks traveling alone to Biskra and Tourggourt, another oasis in the Sahara. Buying the plants was a straightforward matter; he collected 1,000 offshoots of Deglet Noor palms and shipped them to his father in May 1912. Then he returned to California.

By the time Paul got home, the boom had intensified and Fred Popenoe wanted more offshoots. Two months later he sent Paul, twenty-four, and his brother Wilson, twenty, on a trip around the world to collect enough offshoots to surpass the stock of all nurseries in California. Remembering the *Los Angeles Times'* thrilling tales of Frank Meyer's adventures, Wilson was excited to be a real plant explorer. "As Paul and I walked down Calaveras Street [in Altadena] that afternoon in July 1912, bound for Baghdad, I think our mental outlook was very much that of two Boy Scouts headed for their first jamboree," he recalled.[9]

The trip did not turn out to be a Boy Scout jamboree. When they arrived in northern India, on the way to Arabia, Paul was sick with typhoid fever. They kept traveling west despite his illness; by the time the ship arrived in Basra, Paul was almost dead. He survived only because American missionaries nursed him for five weeks until his fever broke. After Paul was able to work again, Wilson caught malaria and dysentery. He recovered after spending weeks in the hospital in Muscat, the capital of Oman, receiving regular quinine injections as he lay in bed listening to the shrill cries of sentinels guarding forts nearby.

Had they known in advance how risky the trip would be, Wilson admitted later, "We probably would have stayed home."[10]

*   *   *

While Wilson was sidelined in the hospital, Paul and Homer Brett, a U.S. consul in Muscat, set off into the interior of the Arabian Peninsula to buy date palms. They traveled sixty miles through the desert under the sultan of Oman's protection in a caravan of eleven of the sultan's best camels, Wilson told his father. They were ambushed twice, yet they escaped unharmed each time.

The assignment was not easy. The Popenoe brothers, who both had fair skin, light hair, and bright blue eyes, must have stood out dramatically in the Mideast. "As we passed through the bazaars [in Basra], merchants would spit on the ground and significantly draw their fingers across their throats," Paul wrote later. "In Baghdad we were chased for a mile by a crowd throwing stones and in one of the seaports of Persia a native suddenly took a shot at us with his rifle, which fortunately missed."[11]

Despite the risks, they did not disappoint their father. The brothers bought 9,000 date palm offshoots in Baghdad and Basra and another 6,000 in Algeria and arranged to have the huge lot—each healthy offshoot stood about three feet tall and weighed thirty pounds—shipped across the Atlantic Ocean. The trees survived the voyage because Wilson Popenoe gave his portable typewriter to the ship's captain in exchange for enough fresh water to keep the palms alive.

During the last leg of the journey from Galveston, Texas, to California, the offshoots filled seventeen refrigerated railroad cars, a load so remarkable that newspapers reported the shipment in detail. Paul Popenoe's separate journey home took long enough for him to write three hundred pages about date palms, a useful guidebook for the many customers his father expected to attract.

"It is hoped that the volume may be of service to the men who are actually engaged in building up an industry that is certain to be one of the largest fruit enterprises in California and Arizona," Paul Popenoe wrote. It was also intended as a sales pitch for his father's latest moneymaking scheme. "Those who get into the date industry at once in the right way can make profits that, for a few years at least, will be extraordinary," he promised.[12]

*   *   *

After Wilson Popenoe finished delivering the date palms to America, he did not know what to do next. Cornell University had offered him a full scholarship, but David Fairchild, the old family friend, advised him against taking it. He lured him to Washington instead. "We were afraid college would ruin Wilson Popenoe," Fairchild explained later. "We wanted to make a plant explorer of him and were afraid that he wouldn't be much good for the U.S.D.A. in the field after he got filled on book learning."[13]

Wilson started his new job immediately. In July 1913, after arriving in New York from his trip to Arabia, he boarded a train to Washington. (He took time off only to go shopping. He spent his last dollar on a new suit because he had ripped the trousers on his old one on the ship.) Fairchild offered him a job as a government plant explorer at a weekly salary of $150, a sum so large that Fred Popenoe scolded his son for accepting it, telling him he should be ashamed about taking so many dollars from taxpayers.

From the beginning of their professional relationship, Fairchild treated Wilson Popenoe as his protégé. "The Chief tried always to help me as 'Uncle Barbour' had helped him," Popenoe said later.[14]

Fairchild made room for him in his family's Maryland home and imbued him with his enthusiasm for foreign plants. Popenoe remembered traveling with his boss one evening on the streetcar from the agriculture department to the Fairchilds' house, which was near Chevy Chase. Popenoe watched Fairchild began a conversation with a stranger and, by the time Popenoe and Fairchild got to their stop, Fairchild had convinced the stranger to plant udo in his garden. (Udo, a Japanese vegetable that Lathrop enjoyed, never caught on as food in America, although high-end nurseries now sell it as an ornamental.)

Fairchild also wasted no time giving Wilson his first assignment: investigate the origins of the single most valuable foreign plant introduction in U.S. history, the navel orange from Brazil.

* * *

Credit for this discovery had always gone to William Saunders, chief horticulturalist and landscape gardener for the United States government. In 1869 Saunders had built the large greenhouse in downtown Washington where Frank Meyer later worked and, like the officials who wrote to Henry Perrine in the 1830s, asked U.S. diplomats around the world to send him plants to test. In 1871 Richard A. Edes, a consul in Brazil, reported that an especially sweet seedless orange grew near Sao Salvador de Bahia. The variety had an odd indentation at one end shaped like a human belly button. The Brazilians gave it the obvious name.

Edes collected cuttings of the orange and sent them to Saunders by ship from Rio de Janeiro to Washington. The trip was long; the voyage was so slow that the cuttings arrived as worthless dried-up sticks, a frequent hazard of plant introduction. Saunders refused to give up. He asked for more and sent detailed packing instructions to Brazil to protect the cuttings. Shortly afterwards a second box arrived in Washington, probably shipped by American Presbyterian missionary Francis J. C. Schneider or his wife. This batch included a few twigs healthy enough to survive; Saunders immediately grafted them onto rootstock he had grown in his greenhouse. By 1873 a dozen little fruit trees were ready for transplant.

Saunders believed that Florida would be the best corner of America to grow navel oranges, so he sent most of them to farmers there. He saved two tiny trees for Eliza Tibbets, his friend and former neighbor in Washington who wanted some, too. Eliza, who had been a lively suffragette and spiritualist when she lived in Washington, had moved out west with her husband, Luther, in 1870 to take their chances as pioneers in southern California.

The year before the Tibbets left Washington, investors had tried and failed to develop a silk farm on thousands of acres near the Santa Ana River. The land was treeless and rocky, but eastern developers had bought 8,735 acres and were seeking "intelligent, industrious and enterprising" settlers to grow semitropical fruit on the site. Luther and Eliza Tibbets were among the first twenty-five families to take the challenge.

Saunders sent his old friend two specimens. The tiny trees traveled by train from Washington to San Francisco and south to Gilroy, where they were loaded onto a stagecoach for the three-day trip to Los Angeles. The Tibbets rode their horse-drawn buckboard wagon sixty-five miles to pick up their package and sixty-five miles to return home in Riverside.

Eliza Tibbets planted the trees in the doorway of their simple house. Because the Tibbets had no running water, local legend says Eliza nurtured them with dirty dishwater. As the plants grew, she shared cuttings with her neighbors. By 1875 the California trees produced sweet seedless oranges with the odd indentation. The Florida trees failed.

On January 22, 1879, one of the Tibbets' neighbors entered the orange in a contest sponsored by the Southern California Horticultural Society. It won first prize and the race was on among settlers to cultivate the fruit. Within a few years, Southern California was covered with orange trees. "It has proved to be, perhaps, the most valuable introduction ever made by the Department of Agriculture in the way of fruits," Saunders wrote in an unpublished account quoted later in a USDA bulletin.[15]

Years later, Beverly Galloway, Fairchild's boss, had some fun with the incident when he wrote, "Trees are like folks. Some come into the world great, some achieve greatness, and some have greatness thrust upon them."[16] Saunders had called the fruit "the Bahia orange" after its original home in Brazil, but the local newspaper's booster publisher changed its name to "the Riverside navel." That angered developers and growers in nearby towns who didn't want to miss out on the new sensation, so the name was changed to the Washington navel. The introduction cost nothing (taxpayers didn't pay for the postage from Brazil), yet it quickly established the orange groves of California. Within a few years the Washington navel had become the most valuable orange in America.

"The founding of miles of orange groves such as the world never saw before is the result of the importation of a single bunch of scions from Bahia, Brazil," Fairchild boasted later.[17]

*   *   *

Ever since Eliza Tibbets and her neighbors introduced the delicious fruit, rumors had circulated that Brazil had other varieties that were even better than the Bahia orange. Fairchild decided that young Wilson Popenoe was the perfect man to find out if the rumors were true.

When he hopped off the train in Washington in July 1913, Wilson was twenty-one, tall, and lanky with slicked-down hair parted in the middle. He had already had more adventures than most Americans; few men his age had traveled around the world and survived a close brush with death in a strange Arabian outpost. He was ready for more.

Fairchild picked two older men to accompany Popenoe on his first expedition. They were Archibald Dixon Shamel, thirty-six, an expert on citrus fruits, and Palemon Howard Dorsett, fifty-one, the soundest gardener in the department. Fairchild was optimistic that the well-balanced team would do well. "Expedition carries the hopes and best wishes of us all," he cabled the explorers when they left America on October 4, 1913. "We feel perfectly confident that you will make it a success."[18]

The trio left New York on the SS *Van Dyke*, the ship that was also carrying former president Theodore Roosevelt to his own expedition on the River of Doubt in the Amazon. As the Van Dyke prepared to depart and a horde of reporters filled the dock to yell questions at Roosevelt, Popenoe found himself standing next to the former president. "Realizing that I was billed to appear in 217,911 moving picture theaters throughout the United States and its island possessions," he wrote in his official report on the trip, "I put on a dignified look and did my best to carry out my part of the program."

*   *   *

Popenoe, Dorsett, and Shamel were an odd team. Popenoe was eager, even sometimes visibly nervous. He read romantic books about the tropics and ogled the sights wherever he went. Shamel was cautious and hesitated to enter a local hut because, as Popenoe told his parents on January 4, 1914, "there might be bugs."

Although he was a gardener, Dorsett was no outdoorsman. Even in the humid jungle he dressed for work like a federal bureaucrat: a dark wool suit with collar, a tie, and a stiff black hat. Dorsett confided to Fairchild that when he was really warm he removed his shirt but left on his jacket. Popenoe said Dorsett's outfit stood out everywhere in Brazil. "His black derby, which he insisted on wearing into the wilds, was always yards ahead of me on the trail," he wrote later.[19]

The three divided their labors: Popenoe, who dressed the part of tropical plant explorer by wearing a white cotton suit, knew enough Portuguese to speak to Brazilians. Dorsett took photographs. Shamel handled all negotiations with government officials. While Dorsett looked like an office worker, Shamel dressed like a United States diplomat; on formal occasions he wore a cutaway suit and silk hat and carried a silver-tipped walking stick he bought in Rio de Janeiro.

The trio spent two months traveling through Bahia, the Brazilian state best known for its orchards, hunting for seedless oranges or other citrus varieties. They failed to find anything even close to the delicious Washington navel; Shamel, the orange expert, determined later that the original fruit was an accident, a natural mutation or bud sport of a Portuguese orange called the selecta. The trio proved that the rumors were false; no better oranges existed in Brazil.

After the team completed its original assignment, Shamel returned to America, leaving Popenoe and Dorsett to set off alone into Brazil's interior to find other plants for America. "This was the era of what I called 'shotgun exploration,'" Popenoe said in a 1963 speech. "We who were in the field always had a major objective or two, but we also had instructions to grab everything in sight which we thought might be of interest for experimental cultivation in the United States."

The two explorers didn't come up with much on this trip, and the few items they found didn't travel well. Shippers mishandled at least two large collections on route to Washington, and all the fruit arrived rotten and useless. One potentially valuable discovery was a forage grass called *Capim gordura* that was an important cattle feed in parts of Brazil. Yet it was difficult to introduce in America. "One of our correspondents wrote us a sarcastic letter saying that the grass grew well but that his cattle refused to eat it," Fairchild recalled later, "so the trouble really was not with the grass but with the cattle."[20]

* * *

The plant hunters found one odd specimen with an unusual name: the jaboticaba, a large purple fruit similar to a grape, that grows directly on the bark of a huge, magnificent tree. When Alec Bell heard the fruit's name—the Tupi Indian word for tortoise that is pronounced zha-bu-ti-ca-ba—he was so taken with the unusual word that he chose it as the name of a hydrofoil boat he invented in Nova Scotia.[21]

The explorers, tanned from the tropical sun despite Dorsett's derby, returned to Washington in April 1914 with about two tons of promising plant material. Fairchild celebrated their arrival with another office party, this one a banquet at a catering hall downtown. About fifty staff members and relatives attended the party, which featured a series of speakers who teased the explorers about their work. The evening's highlight was Fairchild's delivery of a poem he wrote that captured the spirit of plant exploring: "Few man can do what they surely have done / Caught the Jaboticaba without using a gun."[22]

* * *

The Bahia expedition produced nothing of long-term value, yet Fairchild was so satisfied with Popenoe's work that he immediately found another assignment for him. This one involved the avocado, Popenoe's first love. By 1915 Fred Popenoe's 1911 prediction had come true; demand had soared for new varieties in the few years since the Fuerte had survived the sudden frost. Fancy hotels in Los Angeles and San Francisco were paying as much as $1 for one fruit (that's about $23 today). Commercial nurseries clamored for more.

"The avocado with a favorite dressing is pronounced the most delicious of all salads," one promoter wrote later. "Once the taste for the fruit is acquired, the desire to satisfy the palate becomes almost a craving, accompanied by a readiness to pay almost any price for the fruit."[23]

In 1915 Fred Popenoe and other local farmers created the California Avocado Association to capitalize on growing demand. At the group's first meeting, Herbert Webber, the plant breeder who had been Walter Swingle's assistant in Florida before he left to teach at the University of California, proposed that David Fairchild's office send an explorer into the jungle of Guatemala to find the best varieties for Southern California. Who would be better for the assignment than Fred Popenoe's son?

Wilson Popenoe, twenty-four and exploring alone, left New York in August 1916 on the SS *Sixaola*, an 8,000-ton steamship operated by the United Fruit Company. It took a week to reach Puerto Barrios, Guatemala's port.

Popenoe spent the night in the United Fruit's building there and left the next day for Guatemala City, about 185 miles inland. It was a magical trip for a lover of the tropics and its magnificent foliage.

"The narrow-gauge railroad plunged almost immediately into tropical forest, the kind made familiar to the older generation by Rand McNally readers," he wrote later in his autobiography. "Monkeys and alligators were missing, but the gigantic ceiba trees, the immense cohune or manaca palms, and the tangle of lianas were intensely satisfying."

He rode through banana plantations and passed gringos wearing high boots and riding breeches. He entered hot, open country where women constantly fanned themselves against the warm air. After eleven hours, the little train reached the high plain of Guatemala's capital city at dusk.

He woke the next morning to find a young Indian woman at his door offering beautiful orchids for sale for about 15 cents each. "For the moment I did not care whether I found any avocados or not," he wrote in his autobiography. "Guatemala was eminently satisfactory."

Popenoe set up base in Guatemala City and took trips into the countryside that lasted a few weeks. His usual transportation was Starlight, a white horse he bought from an American coffee grower. Popenoe's assistant was a Kekchi Indian, Jose Cabnal, who rode beside him on a mule. The two traveled more than 3,000 miles together on one outing. Their routine was to scout for cultivated avocado trees that looked promising and persuade the owner to let them take cuttings.

It was a rough life under primitive working conditions. Popenoe developed another bad case of malaria and later, like Meyer, he had to worry about political unrest. By summer 1917 Popenoe wanted a gun, despite Fairchild's ditty. "The recent massacre of five white men by the Indians in Santa Maria has made me feel the need of going armed, a thing which I had not previously considered necessary," he told Fairchild on July 10, 1917.

Despite his fears, Popenoe spent most of his time working quietly in avocado groves in peaceful corners of Guatemala. At one point during the expedition he was eating so much fruit that his landlady in Antigua, a mother with two young daughters, almost evicted him because she feared that the avocados, which were reputed to be aphrodisiacal, would drive Popenoe to attack her daughters. By fall 1917 Popenoe had eaten about 1,000 avocados and chosen twenty-three varieties to test in America.

Wilson Popenoe's lifelong dream of plant hunting in the tropics had come true, just in time. When he began his Guatemala expedition, the war in Europe had cooled Washington's enthusiasm for foreign plants and the office could no longer afford to send out teams of explorers. By 1915 Popenoe and Meyer were the only men in the field. "It will be a race between Wilson Popenoe and me to see who will leave to posterity the greater number of introductions," Meyer wrote to Fairchild.[24]

Popenoe returned to Washington by Christmas 1917, in time to receive more professional and personal advice from his boss. Fairchild was working hard to keep his enterprise going. When America entered the First World War in April 1917, Congress began to squeeze his budget, and the American people weren't exhibiting much interest in eating foreign foods. Fairchild knew that the office of plant and seed introduction might shut down if Popenoe quit exploring to, say, get married. As Lathrop had made Fairchild promise he wouldn't marry while he was a plant explorer, Fairchild extracted the same vow from Popenoe.

"He gave me a good strong shot in parting that I was to keep away from the girls," Popenoe told his parents on December 30, 1917. "'If you see you are in danger,' said he, 'turn around and run.' He assured me that if I waited a while I could have anything I wanted. But if I settled down now it would greatly interfere with my career."

Fairchild's insistence that Popenoe stay single was such a joke around the office that one day Howard Dorsett warned that "if I dared to get married before I was 70 years old I lost my job automatically, so there," Popenoe told his parents on January 24, 1918.

Two months after returning to Washington, Popenoe volunteered to use plant exploring as a cover to do intelligence work for the U.S. Navy in Mexico. When he signed up, the recruiting officer in Washington asked him how many languages he spoke. Popenoe answered truthfully, "Five." The officer said, "That's enough. The last feller that came in said he could speak nine. We told him anyone who could speak that many languages wouldn't do anything but talk. We turned him down."[25]

Popenoe's secret mission was to monitor German ship movements off the Mexican coast, but there apparently weren't many, so he spent his time on active duty traveling through the country looking for suspicious activity. There wasn't much of that either, yet Popenoe concluded that he had done good work for the United States. He returned to Washington in January 1919 and started writing about his plant discoveries. His most respected work, *The Manual of Tropical and Subtropical Fruits*, was published in 1920 and remained in print for more than fifty years.

Popenoe returned from Mexico sporting a mustache, a feature that he hoped would attract women. He shrugged off the snide remarks his colleagues made about it. "I hear frequent half-smothered comments on my mustache, but am still bravely retaining it," he wrote his parents on January 24, 1919. "Probably I will get disgusted one of these days and shave it off. It depends on whether or not the girls object to it." He kept the mustache for many weeks, but shaved it off as soon as Marian Fairchild announced that she didn't like it.

David Fairchild, concerned about the future of the office, wrote to Fred Popenoe to praise Wilson and discreetly urge him to remain on the staff and not seek easier money in other lines of work.

"I have felt all along that Wilson was thoroughly devoted to this project of bringing in new plants and had determined to make it his life's work, as I have done, regardless of what might come," Fairchild wrote in January. "I had opportunities early in the game to get out of it and financially would perhaps be better off today had I done so, but there are other things besides money, as you will appreciate, particularly in these days when values are being so tremendously shifted."

Fairchild's advice worked. Wilson Popenoe never followed his father's constant pursuit of easy money, of promising but doomed schemes to get rich. Hunting for plants was enough for him. "I have always said that agricultural exploration is the greatest gambling game in the world," he said later.[26]

Better Babies

Although western gardeners had been experimenting with creating new varieties of plants since the early eighteenth century—a London gardener named Thomas Fairchild (no relation) mixed a carnation with a Sweet William to confect a new flower known as "Fairchild's mule" in 1717—the first U.S. plant hunters confined their work to the straightforward task of looking for plants to grow in America.

As botanists became wiser about plant genetics, however, the explorers' mandate broadened, and they started exploring for foreign breeding material, too. The turning point in their work came at an international conference on hybridization in London in July 1899. David Fairchild, Walter Swingle, and Herbert Webber, Swingle's assistant in Miami, attended the prestigious meeting along with 120 other scientists from around the world. They were the guests of the Royal Horticultural Society, probably the most distinguished botanical organization in the world. The society invited the three young Americans because they had already done groundbreaking work in plant breeding.

Florida oranges probably were their biggest accomplishment. For several years before the gathering, Swingle and Webber had systematically conducted hundreds of trials in their laboratory in muggy Eustis, Florida, trying to create hybrid citrus plants that would survive frost and still taste good. They had scored their biggest success the year before the conference. It was a delicious tangerine-grapefruit combination they called the Minneola tangelo after a nearby town with a Seminole name. Today American nurseries market it as a premium-priced fruit labeled the Honeybell.

The Royal Horticultural conference was an elegant event. On the first day, guests gathered for lunch at Chiswick House, an estate outside London with a large villa surrounded by formal landscaped gardens. Lectures about plant breeding followed the lavish meal. The next day was filled with speeches and a formal banquet for 130 guests at the Hotel Metropole in downtown London. The tables were decorated with fruit and flowers, including hybridized water lilies.

Swingle was thrilled to speak at the dinner. "It is with particular pleasure that I, a cousin from across the sea, rise on this occasion to respond to the toast of 'hybridists,'" he said. He made a gracious tribute to Thomas Fairchild and exulted over what lay ahead for the serious scientists in the room. "It seems to me it is scarcely possible for us to overestimate the future of the hybridist," Swingle said. "Hybridization is the best and noblest branch of horticulture."[1]

The last day included a lunch and garden party in another mansion. At the end of the festivities, an exuberant military band entertained the conference attendees as they boarded a private train to return to London.

One other American attended the conference. He was Willet M. Hays, an agriculture professor at the University of Minnesota, who had become well known in farming circles when he boldly asserted in 1890 that some plants were more valuable for breeding than others. "There are Shakespeares among plants," he said, choosing a phrase that caught the attention of the Royal Horticultural Society.[2]

Although most of the discussion at the London conference concerned flowers and ornamentals, the Americans—always a practical bunch—became most excited about the new techniques' potential to improve fruits and vegetables. Hays returned to the United States determined to establish a national organization to encourage hybridization.

The world's excitement about breeding increased the following year after three European scientists—including Hugo De Vries, Frank Meyer's mentor at the Amsterdam Botanical Garden—each published a paper outlining his independent rediscovery of Gregor Mendel's groundbreaking work.

Mendel, a monk living in a town that is now called Brno in the Czech Republic, spent years experimenting with pea plants in his monastery's garden. He eventually determined with mathematical precision how characteristics pass from one generation of peas to another. In 1866, a local journal published his findings but, in an oversight that still baffles historians, the scientific community ignored the article for thirty-four years.

Mendel was dead by the time experts around the world read the analyses of de Vries and the other scientists and recognized the significance of his work. Mendel's discovery of the principles of genetics was a tremendous breakthrough for everyone, but it was of immediate use to breeders trying to help struggling farmers. "Until Mendel's law was discovered," Webber wrote later, "we had no understanding of why or how this could be done and had to experiment blindly and with no certainty as to whether or not it could be done till success was attained or the experiment was abandoned as a failure."[3]

Henry A. Wallace, who later served as U.S. secretary of agriculture, said genetics saved plant breeders from playing "botanical blind man's bluff."[4]

In 1903, four years after the London meeting, Hays organized a group called the American Breeders' Association to promote hybridization. It began as a small operation; the first session was held at a high school in St. Louis during the larger annual meeting of the American Association for the Advancement of Science. About fifty people showed up, including David Fairchild, who happened to be in St. Louis at the time interviewing brewers about hops.

The association began as an earnest but harmless group. "It has invited the breeders and the students of heredity to associate themselves together for their mutual benefit and for the common good of the country and the world," Hays said at the first meeting.[5] A few months later, Hays moved to Washington to become Secretary Wilson's top deputy. Hays's high-level appointment added credibility to the young organization, which in its early years did little more than hold a few meetings and publish a modest quarterly about farmers' breeding experiences. Fairchild, who at first played only a minor role in the group, claimed that Hays's enthusiasm initially attracted him to the association. "He was a large man and his mind was so full of ideas that they sometimes seemed to tumble over each other when he attempted to express them," Fairchild recalled.[6]

At first, the association confined its activities to plants and animals, but after its first year a radical idea circulated among its members: why not study eugenics, the tantalizing notion that breeding techniques could be used to create superior human beings? The term, which is a modification of the Greek word *eugenes*, was invented in 1883 by Francis Galton, a respected English scientist who also developed the use of fingerprints and devised the first newspaper weather map. Galton used the label for his theories about improving breeding stock, which he said was "equally applicable to men, brutes, and plants."[7]

Scientists in England didn't take eugenics seriously until the rediscovery of Mendel's paper in 1900. Then, because the theory was related to heredity and sounded plausible, eugenics started to attract attention and funding. In 1904 Galton established a eugenics record office in London to study the inheritance of human characteristics. The theory did not catch on in the United States until the breeders' association promoted Galton's ideas.

During its second annual meeting in January 1906 in Lincoln, Nebraska, the association created a separate committee to study eugenics, a step that gave the concept its first stamp of scientific legitimacy in America. This committee's mandate was to investigate human heredity and "to emphasize the value of superior blood and the menace to society of inferior blood."[8] The *New York Times* thought the action was newsworthy enough to warrant a short article on October 31, 1906, under the headline "To Get Better Babies."

At first, the new committee was only one of forty-three panels responsible for an almost comic range of breeding challenges, such as producing better bees, carriage horses, fur animals, fish, carnations, and wild birds. The eugenics committee got off to a slow start; almost a year passed before it had a full slate of members. Another year passed before the panel issued its first report and spelled out its aims. "This committee hopes that the authority and influence of the American Breeders' Association can become a powerful factor in forwarding the study of and interest in eugenics in this country," David Starr Jordan, its chairman, wrote in January 1908.[9]

The committee succeeded in meeting this goal, unfortunately.

*   *   *

By January 1909, the eugenics committee admitted an important new member, Charles Benedict Davenport, a biologist who had been one of the fifty scientists who gathered in the St. Louis high school to organize the association. Davenport began his career as a poultry expert. Hays recruited him to attend the first meeting because he needed animal people to balance out David Fairchild and all the other plant people.[10]

A native of Brooklyn with solid academic credentials, Davenport had always been a busy man. He taught biology at Harvard and the University of Chicago during the school year and ran the Biological Laboratory in Cold Spring Harbor, New York, in the summer. Two weeks before he attended the association's first meeting in 1903, Davenport had received a grant from the Carnegie Institution, then a new charity, to study heredity in animals and insects. Davenport had opened this new research center, which he called the Station for Experimental Evolution, near the Cold Spring Harbor lab on June 11, 1904.

Davenport's new station was a bucolic farm, according to a newspaper article, with "a number of neat enclosures containing demure-looking poultry, dignified goats of a somewhat questionable parentage, judging by their outward appearance; cattle, sheep, a greenhouse, and a 'vivarium' for the accommodation of insects."[11] Alec Bell donated sheep from his estate in Nova Scotia.

Although Davenport had two demanding scientific operations to run, he wanted another. At the end of 1909, using his position as a member of the new eugenics committee, he prodded the breeders' association into elevating its status from being one of dozens of committees to becoming one of its three major fields of study. In the beginning of 1910 the members voted overwhelmingly—495 in favor, only 5 against—to make eugenics equal to plant and animal breeding. Davenport assured them their decision was sound. "The association of research in eugenics with the American Breeders' Association is a source of dignity and safety," Davenport wrote in January 1910. "It recognizes that in respect to heredity man's nature follows the laws of the rest of the organic world. It recognizes that human heredity is a subject of study for practical ends; the ends, namely, of race improvement." Davenport promised the new section's efforts would be "hard-headed, critical, and practical."[12]

Mendel and other early geneticists had proven that certain physical characteristics of plants, animals, and humans are inherited. Davenport, however, pushed their findings beyond physical to mental characteristics and concluded that humans' personality traits could be inherited as well. To prove his theory, Davenport devised a family history form that listed thirty-five characteristics—such as stammering and peculiar finger movements—that he believed could be inherited. To gather proof, he distributed 5,000 of these questionnaires to the 1,000 or so members of the breeders' association and their colleagues and asked them to answer questions about themselves and their relatives. Responses were sparse and slow, however, and Davenport decided he needed a bigger, more elaborate operation than the volunteer network of association members.

Looking for money, he consulted *Who's Who* and made a list of all the rich people on Long Island.[13] He zeroed in on Mary Harriman, the recently widowed spouse of Edward H. Harriman, a railroad magnate. Davenport, like Willet Hays, was an exuberant man with a persuasive manner. "Few in this field have wielded the power he won, primarily by virtue of the infectious quality of his enthusiasm," a longtime associate wrote after Davenport's death. "Flashes of this shot out in all directions, started uncountable new activities and organizations and gave him control of large funds."[14]

At lunch at Harriman's home on February 16, 1910, Davenport laid out his plans to open a third center in Cold Spring Harbor called the Eugenics Record Office to collect and analyze information about hundreds of human characteristics. He wanted it to serve as a clearinghouse for facts about bloodlines and family traits to help people choose between fit and unfit marriage partners. Later this mission was described as essential to eugenicists' ultimate goal: the elimination of inferior people.

Davenport was indeed persuasive. Harriman agreed not only to contribute to his cause but also to finance the entire operation. Her support eventually totaled more than $500,000 (or more than $11 million in current dollars). Harriman's support thrilled Davenport; after he returned home from their first lunch he wrote in his diary that February 16, 1910, was "a red letter day for humanity!"[15]

* * *

In the beginning David Fairchild was a passive member of the association. But changes occurred in 1913 that carried him from plant introduction to plant breeding and straight into the maelstrom of American eugenics.

The shift was triggered by the election of Woodrow Wilson, a Democrat, in the 1912 presidential election. When Wilson took office in March 1913, he removed James Wilson, a Republican, as secretary of agriculture. Willet Hays left as assistant secretary at the same time and also resigned as president of the American Breeders' Association.

Without consulting Fairchild, Hays appointed him as his successor at the association. Fairchild said he tried to get out of the job, but Orator Cook, his colleague at the office, persuaded him to accept the responsibility. "You can't kick a baby off the doorstep, Fairchild," Cook said. "You'll have to take it in."[16]

Fairchild quickly learned that the association was broke. Correspondence showed it owed bills totaling $3,290.23 but had assets of only $2,510.[17] Rather than scale back its programs, Fairchild wanted to transform the association into a more ambitious operation with a monthly magazine that would attract more readers and more money. The way to do that, he decided, was to capitalize on the growing interest in eugenics. "There is a general recognition among members of this association that the need is urgent at the present time for active and scientific leadership of the rapidly growing eugenics movement, lest it be detached from its proper base of genetics and be captured by sentimentalists and propagandists with slight knowledge of its biological foundation," Fairchild explained in a statement.

Although this plan came early in the history of eugenics in America, there was already heated debate about the validity of eugenics and about how to apply the concept to society. Since at least 1908, Alec Bell—a man always trying to improve human life—had made a careful distinction between positive and negative eugenics and warned against enacting, for example, any marriage laws that would restrict any individual's pursuit of happiness.

In July 1913 Bell attended the first International Eugenics Congress in London at Davenport's invitation. There he heard at least one prominent scientist, Reginald Crundell Punnett, a British geneticist, warn enthusiasts against going too far. "Except for a very few cases," Punnett said, "our knowledge of heredity in man at present is far too slight and far too uncertain to base legislation upon."[18]

FIGURE 14. Paul Popenoe as a young man. Courtesy of the Paul Popenoe Collection, American Heritage Center, University of Wyoming.

Despite this uncertainty about the legitimacy of eugenics, Fairchild relied on Charles Davenport and his supporters to help him save the American Breeders' Association.

Fairchild's first act was to shake up the association's unexciting biannual publication by broadening its mandate to include discussions of new ideas about breeding. He also insisted that writers submit photographs to illustrate their articles. He persuaded Barbour Lathrop to donate funds to print more frequent issues. He enlisted Bell to contribute ideas, articles, and, of course, the credibility and attention generated by his famous name. He changed the organization's title to the American Genetic Association to emphasize its interest in eugenics.

Decorum was the reason for the change, Fairchild claimed in his autobiography. "In those days the terms 'breeder' and 'breeding' could not be used indiscriminately in mixed audiences," he wrote. "The name breeders would have to be eliminated because it smacked too much of the barnyard."[19] He chose the *Journal of Heredity* as the publication's new name.

Probably remembering how Bell invigorated the *National Geographic Magazine* by hiring Gilbert Grosvenor as editor, Fairchild wanted an eager full-time staff member to produce the reenergized journal. An expert from Cornell University was supposed to take the job, but he wanted more than the $1,200 a year salary Fairchild could afford, so Fairchild appointed Paul Popenoe as editor.

Like his brother Wilson, Paul, a twenty-five-year-old plant lover, had returned from the Mideast without a job. He was available to start work immediately, but he was an unlikely and unqualified choice. His education had consisted of two years at Occidental College and a brief stint at Stanford University, where he took a biology course from David Starr Jordan, an early eugenicist. This background had not prepared Popenoe to edit scientific articles on controversial subjects.

He was, however, an enthusiastic man with a clear writing style who did what he was told. Fairchild explained his mission succinctly. "The idea is to show that plants and animals obey the same laws of heredity and that those laws are the ones that govern *Homo sapiens* as well," Popenoe told his parents.

Fairchild may have hired a skilled promoter, but he ended up creating a true believer.

* * *

Even before his appointment became official, Popenoe sought Davenport's assistance. "We want to give eugenics the attention it deserves and must depend largely on you to act as intermediary in this line," he wrote on August 27, 1913. Six days later, he asked Davenport to read a proposed article to check its accuracy so "the magazine may get nothing that will not stand the scrutiny of the scientific world."

By then Davenport was a celebrity. He was running three separate operations in Cold Spring Harbor: the Eugenics Record Office to study humans, the Station for Evolutionary Experiments to study animals and insects, and the Biological Laboratory to teach science to summer students. He had published a popular book on eugenics and had led the United States delegation to the 1912 London conference. America's most prominent philanthropists had supplied him with plenty of money to finance his work.

Aside from obtaining regular stipends from Mary Harriman and the Carnegie Institution, Davenport had persuaded John D. Rockefeller to pay for training field workers to collect information about families. The workers, who were mostly young single women, were so enthusiastic about their work that they wrote a song about it. "We are eu-gen-nists so gay /

and we have no time to play / serious we have to be / working for posterity," they sang at the Eugenics Record Office.[20]

* * *

Due largely to Davenport's enthusiasm, eugenics in America had evolved from an interesting, mostly harmless theory into the adopted cause of racists seeking scientific justification for their bigotry. They wanted to get rid of poor, troubled, and foreign people, and Davenport, a respected scientist, had provided a way to do it when he asserted that criminality, feeble-mindedness, and pauperism were inherited mental characteristics.

James Watson, the Nobel Prize–winning biologist who took over the Cold Spring Harbor Laboratory long after Davenport had been discredited, summed up the damage done by eugenics in the lab's 1996 annual report: "Through propagating such racial and religious prejudices as scientific truths, the American eugenics movement was, in effect, an important ally of the ruling classes, many of whose privileges inevitably came through treating those less fortunate as inherently unequal."[21]

As the journal's editor, Popenoe wrote or edited a dozen articles a month and traveled around the country speaking on behalf of eugenics. Four months into the job, he attended the first conference of the Race Betterment Foundation in Battle Creek, Michigan, an association started by John Harvey Kellogg, the brother of the owner of Kellogg Cereals. At the meeting Popenoe heard Harry Laughlin, Davenport's deputy as director of the Eugenics Record Office, rally the crowd to action. "To purify the breeding stock of the race at all costs is the slogan of eugenics," said Laughlin, who had been a biology teacher in Kirksville, Missouri, before Davenport hired him.[22]

Laughlin was leading a call for forced sterilization of mental patients, criminals, and other people he deemed to be "degenerate"—one of the eugenicists' favorite words—so they could not pass their personality traits on to children. In 1910 Davenport had estimated between 3 and 4 percent of Americans were unfit to become parents, but by the 1914 Race Betterment conference Laughlin's estimate had multiplied. "In these calculations it is assumed that the lowest 10 percent of the human stock are so meagerly endowed by nature that their perpetuations would constitute a

FIGURE 15. Harry Laughlin, left, and Charles Davenport outside the Eugenic Records Office in Cold Spring Harbor, New York, ca. 1913. Photo from the Harry H. Laughlin Papers, Pickler Memorial Library Special Collections, by permission of Truman State University.

social menace," he said.[23] That meant he wanted to sterilize about 10 million Americans, a move that, in the words of Laughlin's biographer Frances Janet Hassencahl, "tends to equate the sterilization of a human with the spaying of a dog."[24]

Paul Popenoe liked what he heard, and he left Michigan convinced that Laughlin's approach was correct. The *Journal of Heredity*'s new editor began writing articles about the need to operate on mental patients and prison inmates to prevent them from having children, with or without their consent. Popenoe, a college dropout without rigorous training, asserted in his articles that eugenics was sound science. "Galton himself has laid down the principle that the most important thing to breed for was energy," Popenoe wrote in a book review published in October 1914. "To this we may add good health, longevity, intellect and morality; for few will be found to deny that these characters form desirable attributes and that they are heritable, even though we cannot show the exact method of their inheritance."

Not everyone approved of the journal's new direction. Early on, Galton's associates in London had begun to question Davenport and Laughlin's methodology. And Thomas Hunt Morgan, a biologist at Columbia University who had been another original organizer of the breeders' association, was outraged. In January 1915 he resigned as chairman of the American Genetic Association's committee on animal breeding because he no longer wanted his name published in the magazine.

"For some time I have been entirely out of sympathy with their method of procedure," Morgan told Davenport on January 18, 1915. "The pretentious title, for one thing, the reckless statements, and the unreliability of a good deal that is said in the journal are perhaps sufficient reasons for not wishing to appear as an active member of their proceedings by having one's name appear on the journal." Morgan added, "I think it is just as well for some of us to set a better standard and not appear as participators in the show." In 1933 Morgan, one of America's most respected early geneticists, received a Nobel Prize in physiology or medicine for discovering chromosomes' role in heredity.

Popenoe didn't listen to Morgan, however. He kept promoting Davenport and Laughlin's ideas and joined the campaign for forced sterilization. Popenoe shared their conviction that degenerate behavior was inherited, and he dismissed as sentimental any arguments against that position. "We eugenists have a stronger faith because it is based on things that are seen and that can even be measured," he said in a speech to the Young Men's Christian Association in 1915. "We think we can prove that it is, on the whole, man who makes the environment, not the environment which makes the man."

* * *

Fairchild was preoccupied during the time Popenoe edited the magazine. Turmoil at the department and health problems at home, as well as having to direct extended plant expeditions, gave him little time to supervise Popenoe. "David is trying to do too much," Lathrop complained to Marian Fairchild as early as September 12, 1913. "In the slang of the West, he is a natural born worker and oughtn't to have too many irons in the fire."

Fairchild did make certain, however, that the journal published articles about plants and exploration as well as eugenics, sterilization, and arguments in favor of tighter immigration rules. Paul Popenoe wrote long articles about date palms, Wilson Popenoe reported on his hunt for the Jaboticaba, and Frank Meyer even contributed a short piece about breeding stags for their antlers in China, for example.

After about a year under the new leadership, Fairchild said he was proud of his new editor. "Popenoe has secured 1,400 new members to our genetic association in 14 months," he boasted to Mabel Bell, Marian's mother, on December 3, 1914. "He is an exceptional fellow." Popenoe remained editor until November 1917 when he was drafted into the U.S. Army. After that, he continued to write articles about eugenics for the journal.

Despite Popenoe's extensive coverage of the subject, Fairchild's scheme to save this publication by promoting the controversial idea didn't work, and its financial problems increased. In 1920, after the cost of paper rose and the printer raised his prices twice, Fairchild went directly to Davenport and bluntly asked for money. "I shall be appreciative of anything you can do to show Mrs. Harriman that we are trying our best to keep the fires burning during these trying times," Fairchild wrote on May 29, 1920.

A month later Davenport reported that Harriman wouldn't help the journal. He suggested instead that Fairchild raise membership fees from two to three dollars a year. Two years later, Davenport admitted that his pledge to bring in as many as 2,000 new members had failed.[25]

The *Journal of Heredity*'s preoccupation with eugenics faded after that, but Paul Popenoe's zeal did not. Wilson Popenoe had been telling their parents for years that Paul had become a convert to Davenport's cause. Paul "is getting to be a big eugenist," he told them on March 13, 1915.

The brothers were very close. They often lived together in a rooming house in downtown Washington where they spread their mattresses on the floor and decorated their rooms with rugs, cushions and other souvenirs from Arabia. They borrowed each other's clothes and searched the city for cheap, tasty meals. In their spare time they chased young women. Wilson was a lively, good-humored young man who enjoyed dancing and flirting, but Paul was more serious. Even in his private life he was on a mission to improve the human race. "When Paul will cross the street just to size up a female passerby or slow down at the corner to wait for one to come up so he can take a look at her, why I say it is time to call a halt," Wilson told their parents in an undated letter written at this time. "He is absolutely insane on the anthropological business (as he calls it)." In another letter Wilson complained that his brother's attitude was spoiling his own fun. "Paul has got me so scared of going with the hoi-polloi that I have to think twice now before deciding whether I am losing caste by speaking to someone I met on the street," Wilson said.

After he left the journal, Paul wrote and assembled articles—including reprints from the *Journal of Heredity*—into *Applied Eugenics*, a work that became the standard college textbook about the would-be science. Paul's co-author was Roswell Hill Johnson, an associate of Davenport's at the Eugenics Record Office. As payment for the work, Johnson offered Paul a thousand dollars or his name first on the cover. Paul chose top billing.

After the First World War ended, Paul Popenoe moved to New York City and finally found a young woman he considered suitable. She was eighteen; he was thirty-one. Her name was Betty Lee Stankovitch; she was born in Illinois to a family of Croatian background. After her family moved to New York, she left school after ninth grade to study rhythmic dancing at Carnegie Hall under Florence Fleming Noyes, a teacher who, like Isadora Duncan, dressed her disciples in flowing silk dresses and diaphanous scarves in the Grecian style.

On their first date, Popenoe was not romantic. "He wrung me dry of information about myself in an hour, then looked at his watch and left," Betty said years later.[26] Their subsequent dates must have been better because, after knowing each other for only about six months, they got married and moved to the California desert so Paul could take over the family business of selling date palms.

News of the alliance shocked David Fairchild. "Paul has paralyzed us all with his engagement to the danseuse, if that is the way you spell it," he wrote to Wilson Popenoe on April 18, 1920. "Do you know her? Is she

FIGURE 16. Betty Popenoe as a young woman. Courtesy of the Paul Popenoe Collection, American Heritage Center, University of Wyoming.

pretty and will she stick to him through the thick and thin days of date growing?"

Their new home was a shack near Coachella, a town that lies so far below sea level that one of its first newspapers was named the *Submarine* (and a current magazine is called the *Periscope.*) The shack was not finished when they arrived: it had four walls but no roof.

Betty Popenoe, a teenager with little education, had trouble adjusting to her new life in the desert. When she met Paul Popenoe, she was living in a boarding house on Park Avenue and 81st Street on Manhattan's Upper East Side. A short time at a summer camp in Maine had been her only experience outdoors.

For the former rhythmic dancer, "the move to the desert was a wrenching experience," one of their sons wrote later. "Pursuing the clearly etched female gender roles of her times, she desperately sought to be the good wife and mother, but the social isolation, the intense heat, the inadequate housing, the sandstorms, the scorpions, and the need to carry a pistol to kill rattlesnakes were terribly trying to her."[27] She stuck it out, however, and proved to be the good breeding stock her husband wanted after she gave birth to four sons.

Meanwhile, Paul Popenoe, like his father, tried to cash in on the potential profits from date palms, but it was hard. After five years of tough living in the desert, Popenoe sold his stake in the Coachella plantation and returned to eugenics to make his living.

About 1926 Ezra Seymour Gosney, a wealthy citrus farmer in Pasadena who also agreed with Harry Laughlin's ideas, hired Popenoe to evaluate California's sterilization law. Under that statute, which was enacted in 1909, officials had already forcibly operated on approximately 6,000 prisoners and mental hospital patients. In 1928, the two men started a new eugenic organization called the Human Betterment Foundation to promote sterilization throughout America. The following year they issued a report that pronounced California's law a big success.

In the 1930s, the organization worked closely with scientists in Nazi Germany, although the Nazis preferred the term *racial hygiene* to the word *eugenics*. On January 1, 1934, the Nazi government enacted the German Sterilization Law, a national statute based on recommendations from Paul Popenoe and Harry Laughlin. Popenoe reported the news in the *Journal of Heredity* in an approving article.

Under provisions of the new law, the German government intended to establish 1,700 tribunals called "Eugenics Courts" to examine 400,000 people to determine if they should be sterilized. The Nazis seem "to be proceeding toward a policy that will accord with the best thought of eugenists in all civilized countries," Popenoe wrote. "In any case, the present German government has given the first example in modern times of an administration based frankly and determinedly on the principles of eugenics."[28] Earlier that year Popenoe predicted that other nations would soon follow Germany's lead. Sterilization "seems likely to take a place in the humanitarian program of every civilized country in the near future," he wrote in the *Journal*, adding more praise for California's program, which remained the largest in the United States.[29]

Popenoe's enthusiasm ignored the scientific community's growing disgust with eugenics. Hermann J. Muller, an American geneticist and another future Nobel laureate, wrote in 1935 that eugenics had become "hopelessly perverted" into a pseudoscientific façade for "advocates of race and class prejudice, defenders of vested interests of church and state, Fascists, Hitlerites, and reactionaries generally."[30]

By January 1, 1938, 12,180 people in California had been forcibly sterilized, more than double the total when Popenoe first became involved in the cause. Despite the growing outrage over eugenics, Popenoe and Gosney's organization remained enthusiastic. "Eugenic sterilization primarily is applied by the state or with its sanction to persons who would be likely to produce defective children," read "Human Sterilization Today," a pamphlet issued by the Human Betterment Foundation about this time. "It protects such persons, their potential children, the state and prosperity. Such persons do not have the intelligence, the foresight or the self-control to handle contraceptives successfully nor the ability to care for children intelligently."

In 1942 Gosney died, and the foundation closed. By then America was at war with Germany, and the horrible consequences of the Nazis' eugenic laws were becoming known. After the war, Popenoe corresponded briefly with an old Nazi friend named Otmar Freiherr von Verscheuer, director of a eugenics institute in Germany and the boss of Joseph Mengele, the notorious doctor at Auschwitz. In one letter Popenoe's tone was friendly, but he claimed to be ignorant of Nazi practices. "None of us knows anything about what was going on in Germany from about 1939 onwards," Popenoe told Verscheuer in 1946.

By then Popenoe had stopped writing about eugenics and forced sterilization. He had switched his attentions to a milder form of social engineering: keeping unhappily married couples together. Popenoe, who claimed to have invented marriage counseling, became a minor celebrity in the 1950s through his television appearances and a long-running advice column in the *Ladies' Home Journal* called "Can This Marriage Be Saved?" Divorce, not degenerate human beings, became Paul Popenoe's new obsession. And it apparently gave him the financial success he—and his father—had sought for so long.

The *Journal of Heredity* has survived as a solid academic journal covering genetics and related scientific subjects. To replace Paul Popenoe, Fairchild gave the editor's job to Robert Cook, the son of Orator Cook, the man who persuaded him to take control of the American Breeders' Association in the first place.

13

A Chinese Wall

When David Fairchild married Marian Bell, the couple had only one attendant at their wedding. That honor belonged to Charles Marlatt.

It was simple springtime ceremony, but the elegant setting showed how far the two men had traveled since they were students in Kansas. The marriage was performed on the front lawn between two large trees at Twin Oaks, the huge estate of Marian Bell's grandparents in Washington's Cleveland Park section. Gardiner Hubbard and his wife, Gertrude, had bought the forty-five-acre property in 1888 as a summer place—they spent each winter a few miles away in a mansion near Dupont Circle—and transformed the original house, a simple, authentic colonial home, into a twenty-six-room Colonial Revival mansion, surrounded by perfectly groomed gardens. It was reputed to be the most beautiful private park in Washington.

Twin Oaks was a long way from the small stone farmhouse of Marlatt's childhood in Kansas. His parents had been among Manhattan's earliest and most idealistic settlers. His father was George Washington Marlatt, an itinerant Methodist minister and strong abolitionist who had arrived in Kansas on foot from Indiana in 1856. He helped organize the community's first educational institution, which despite its name—Blue Mont College—was a primary and secondary school, and he served as its first principal. Marlatt hired Julia Bailey, a young woman from Connecticut, as the assistant principal and, a few years later, married her.

FIGURE 17. Charles Lester Marlatt as a young man. Photo by Francis Benjamin Johnston from the U.S. Library of Congress, item no. 93512848.

After struggling for five years to educate the small town's children, Marlatt and his associates donated Blue Mont's buildings and grounds to the state government to serve as Kansas State College's first campus.

Although Washington Marlatt began, like David Fairchild's father, as a minister and educator, he turned to farming full time about the time Charles was born. Julia, who raised their five children, became a passionate collector of insects while living on the farm in Manhattan. She often took young Charles to entomology lectures with her.[1]

At Kansas State, Charles Marlatt, who was six years older than Fairchild, was a student of Edwin A. Popenoe, Wilson and Paul Popenoe's uncle and an entomologist who served as chairman of the department of botany and horticulture. After he graduated from college, Marlatt stayed on campus to work as Popenoe's assistant. Long before the wedding, Marlatt had unknowingly influenced Fairchild's future when he arranged Alfred Russel Wallace's visit to the school to discuss his adventures in Java.

Charles Marlatt was a gifted illustrator, then a valuable skill for an entomologist. In 1889, two years after Wallace visited Kansas, Edwin Popenoe asked Charles Valentine Riley, the well-known entomologist, to add Marlatt to his expanding staff at the U.S. Department of Agriculture. Unlike Fairchild and Walter Swingle, who joined the department as botanists, Marlatt's job was in the entomology section. The three young Kansans often worked together even though they specialized in different fields of biology, branches of science whose interests sometimes clashed dramatically.

*   *   *

Marlatt was twenty-six when he arrived in Washington on January 1, 1889. He was tall man with classic features, light brown eyes, and dark hair. Determined to succeed, he snagged an invitation to a New Year's Day reception at the White House on his first afternoon in town.[2]

Although he was supposed to confine his work to illustrating other scientists' articles about insects, Marlatt quickly showed his superiors that he would be more valuable as a scientist then as an artist. (His talent was not wasted, however; years later he put his remarkable skills to work when he decorated the woodwork in his Washington, D.C., home with engravings of insects, fish, butterflies, and other creatures, each drawn to scale.)

When Marlatt started at the agriculture department, his boss gave him several articles in French about wine making and told him to write a paper on the subject for American farmers. Marlatt knew no French, but he managed to learn enough, fast. He hired a native French speaker to eat dinner with him every night, and he read and reread a French novel with short sentences and paragraphs. Marlatt methodically translated each word until he could read the book and understand enough French to translate the winemaking articles into English. Next he wrote a fifty-eight-page instruction pamphlet.

Spirits were high in 1889 at the entomology division, a unit with only about eighteen employees. Riley, the boss, was a rarity: a respected scientist who was also a celebrity, at least in agricultural circles. Born in London in 1843, he had studied art and natural history in Europe. He moved to Illinois to work as a farmhand when he was a teenager. There he taught himself entomology to try to understand why insects destroyed crops on the farm. Later he became a reporter for an influential weekly newspaper, the *Prairie Farmer,* where he wrote articles about the insect woes that bedeviled American farmers, especially the terrifying swarms of locusts that ravaged Kansas and the rest of the Great Plains in the 1870s.

Riley tried to allay panicked farmers' fears. The attacks, he wrote, were a scientific phenomenon that would end naturally as the insects worked through their normal life cycle. In the meantime, Riley comforted farmers with suggestions that they capture the locusts and eat them for protein. (He recommended frying them in oil, with lots of salt and pepper, or cooking them in broth, like crawfish, for soup.) Government officials were so impressed by Riley's stories that they hired him away from the newspaper to serve as Missouri's state entomologist.

While locusts were devouring crops in the Midwest, a different pest was attacking grapevines in Missouri, a state that German immigrants had turned into an important winemaking center. In his search for a way to eliminate the pest in Missouri, Riley corresponded with Jules-Émile Planchon, a French botanist who was trying to understand phylloxera, a disease that was killing grapevines in France. The relentless, mysterious death of the wine industry was devastating the country. Riley figured out that the insect causing phylloxera was a variety of a common oak tree pest, but the insect didn't attack domestic grapevines because it had a natural enemy in America. He recommended that French vintners grow their grapes on American rootstock, a suggestion that saved France's wine industry. In gratitude for his help in 1844, the French government gave him its highest civilian honor, the title of Chevalier de la Legion d'Honneur.

Even before he earned that award, Riley had an impressive reputation with an ego to match and a dramatic personal style. "He looked more like a poet or an artist, or possibly an actor, than a man of science," observed Leland O. Howard, Riley's longtime assistant, who joined the agency in 1878.[3]

In 1888, the year before Marlatt arrived, Riley again used his good sense and daring to help American farmers. A disease called cottony cushion scale was attacking citrus trees in California. Although apparently carried innocently on blossoms in bouquets, it was threatening the survival even of the navel orange orchards of southern California. Riley understood that Australian citrus trees suffered a similar problem but withstood it because the country had its own natural enemy there, a ladybird beetle called *Vedalia cardinalis* that devoured the scale before it destroyed the trees.

Riley loved to travel, but his lavish spending of government money infuriated officials in Washington. Congress grounded him by attaching a rider to the agriculture department's budget law that barred any employee from spending government money outside the United States. Always determined and sometimes devious, Riley overcame this obstacle by persuading the U.S. Department of State to pay for an entomologist on his staff, Albert Koebele, to attend a scientific convention in Melbourne. In Australia Koebele collected ladybird beetles and carried them to America, after spending only State Department funds, not Agriculture Department money. The bugs immediately went to work and solved the farmers' problem. They eradicated the disease in California, allowing the citrus trees to thrive. Riley was so proud of his accomplishment that he named his daughter Cathryn Vedalia Riley.

Despite his ingenuity, Riley's excesses and tendency toward self-promotion cost him his job. In 1894, after President Grover Cleveland named J. Sterling Morton as the new agriculture secretary, Riley lobbied for promotion from chief of the entomology division to assistant secretary of the department. He threatened to resign if he didn't get the job. Yet, to Riley's surprise, Morton passed him over for Charles W. Dabney. Two months later, Morton accepted Riley's resignation. The next year Riley, fifty-one, died from injuries suffered in a bicycle accident in Washington.

Replacing Riley was tricky. The logical candidate was Leland Howard, who had served for more than a decade as Riley's top assistant, a job made difficult by their different personalities and backgrounds.

Howard, unlike Riley, did not start out by getting his hands dirty as a farmhand in Illinois. He learned about insects the refined way at Cornell University, where he graduated with a degree in entomology, and at Georgetown University, where he took graduate courses after he moved to Washington. One of Howard's jobs at the agency was to research, write, and publish scientific articles—under Riley's byline. Howard said the practice was common at the time, but he didn't like it. "Assistants accepted the situation," he recalled, "not because it was right, but because there was nothing they could do about it."[4]

Howard said Riley treated him more like a clerk than a fellow scientist. When Riley resigned from the agency in 1894, Howard expected that Assistant Secretary Dabney would promote him to run the entomology division. But Dabney had disliked Riley, an attitude that soured him on Howard, too. "He feared, since I had been his 'loyal' assistant for so long, that I was probably the same type of man," Howard wrote later.[5]

Other scientists convinced Dabney that Howard was not like Riley, and in June 1894 he promoted him. Howard made certain he had his own truly loyal deputy. He picked Charles Marlatt.

At this time, scientists were worried about a new disease that was ravaging apple, pear, and other stone fruit trees along the Eastern Seaboard. The affliction was called San Jose scale because botanists had spotted it first in the 1870s in San Jose, California. American scientists blamed its introduction on an insect carried on plants from Asia, although they didn't know any details. For decades, officials in Washington had pretty much ignored the disease because it was confined to the West Coast. In August 1893, however, it turned up on trees in Charlottesville, Virginia, and kept spreading. As the infection of San Jose scale moved inexorably along the East Coast, agriculture officials began to panic. If the disease destroyed all of America's fruit trees, the potential losses were enormous.

Scientists understood, of course, that foreign plants could introduce foreign pests and diseases to America. Swingle, always the young man with brilliant foresight, was one of the first botanists to speak publicly about the problem. In April 1893, he told the Florida State Horticultural Society that reducing this risk was one reason to establish a private company to inspect foreign citrus plants. Fairchild had stressed the same point in the first article he wrote outlining the office of seed and plant introduction's goals in 1898. "The danger of inadvertently introducing new weeds, noxious insects, or dangerous parasitic fungi is one which should not be underestimated," Fairchild stressed in *Systematic Plant Introduction*. "This danger, however, can only be reduced to a minimum by narrowing the avenues of plant introduction and subjecting every importation to careful inspection and disinfection."

The scientific community generally agreed with Fairchild that inspection and fumigation were effective ways to protect American crops, combined whenever possible with the introduction of pests' natural enemies. Howard, however, didn't agree with most scientists and took a tougher stance. As soon as he gained control of the office of entomology, Howard launched a campaign to keep foreign plants out of the United States. It was a significant change in policy, one that eventually propelled the arcane study of bugs into the thickets of American politics and international affairs.

Howard implemented the new policy discreetly. A few months into his new job, he gave a speech about the risks of bringing in plants from Mexico. Next he distributed booklets to farmers that itemized each state's rules on plant imports. Finally he began a drive to outlaw foreign plants.

In March 1897 Leland Howard and Beverly Galloway organized a national convention to discuss the best way to block insect pests and plant diseases. They invited state officials, commercial nurserymen who imported plants, and scientists from the agriculture department to attend a two-day session at the Ebbitt House in Washington. The convention opened the day after President McKinley's inauguration.

Galloway, Fairchild's mentor, was hopeful and positive when he addressed the group. He viewed the increase in pests as a sign of progress, a mark of America's growing diversity. "Where an insect or fungus had one chance a hundred years ago to wax strong and spread," he said, "it has now a thousand chances, for unbroken orchards and vineyards and millions of nursery trees cover the country where then only wild plants grew."[6]

Galloway represented the position of most scientists when he argued that enacting a federal law would be useless because foreign insects and diseases would enter the United States no matter what Congress did. "It would be manifestly as impossible to control such enemies by legislation as it would be to control the dust of the air or the wind that wafts it from place to place," Galloway said. "The greater portion of our plant diseases and insect pests cannot be reached by legislation. They are governed by natural laws, and it is to these that we should turn our attention."[7]

Galloway's proposed solution was to grant the agriculture department's scientists special powers to mobilize quickly in case of an emergency. He wanted the agency to be able to appoint inspectors, order quarantines, destroy plants, and even overrule state officials whenever anyone discovered signs of disease. The department, he argued, already had similar authority over livestock.

Howard disagreed. He insisted that the United States urgently needed laws to restrict and, if necessary, block foreign plants, although he didn't push for a total ban on imports at the session. "Fully one half the principal injurious insects in the United States are of foreign origin and have at one time or another been accidentally imported into this country," Howard said.

Nine months after this convention, a bill drafted by Howard to empower the department to inspect foreign plants was introduced in the House of Representatives, but members ignored it. A few weeks later, foreign countries—whose leaders were frightened that U.S. exports would carry the San Jose scale and infect their own fruit trees—started to pass laws barring American plants. Germany, Canada, and other important trading partners were soon blocked to American farmers. Even Java banned U.S. plants.

Still, scientists in Washington—including Charles Marlatt, Howard's top assistant—insisted that barring foreign imports would be futile. On August 18, 1899, Marlatt delivered an address to the Association of Economic Entomologists in Columbus, Ohio, about the uselessness of trying to control nature by what he called humans' "puny efforts." Howard was in the audience and heard his deputy challenge his own position on this controversial subject. "If the principle of protection from foreign pests were followed out legitimately by all countries, it would practically stop the commerce of the world," Marlatt said. "Either we must build a Chinese wall and live entirely apart from the rest of mankind or make up our minds ultimately to be common possessors of the evils as well as the benefits of all the world."

Marlatt, who had worked for Riley for several years, said the best way to fight the disease was to introduce the San Jose scale's natural enemy—if the insect could be found. Two years later Marlatt set off to look for it himself.

* * *

In 1901 he took a one-year leave from the department to hunt for the bug that could eradicate San Jose scale. Marlatt had married Florence Lathrop Brown of Boston (possibly a distant relative of Barbour Lathrop) five years earlier, so the trip was part delayed honeymoon, part insect expedition.

For the mission to succeed, Marlatt would have to locate the scale's original home, but he had only one clue to follow. Gardeners discovered the scale first in an orchard in California owned by a man named James Lick, but Lick was dead and no one knew exactly where he had gotten his trees. Experts believed the source was somewhere in Asia.

Marlatt and his wife arrived in Japan in April 1901 just in time to participate in the huge annual cherry blossom festival. The Marlatts attended a series of parties to celebrate the trees in Tokyo and took a leisurely drive though the park in Ueno, which was lined for four or five miles with splendid cherry trees. Marlatt, ever the entomologist, remarked in his account of the trip not about the trees' beauty but about a slight insect infestation he noticed during the drive. At tea one afternoon, he also showed, however, that his interest in insects was limited to looking at them: he declined his host's offer to taste sugared examples. "Some candied or glaceed locusts, crickets, dragon-fly pupae, and other insects were presented," he wrote in An Entomologist's Quest. "We sampled these by sight only."[8]

Another day the couple attended the emperor's annual cherry blossom garden party in formal dress—Marlatt wore a top hat and frock coat—but as usual he was on the lookout for insects. The emperor's cherry trees were hundreds of years old and "had an unbroken record from the earliest times to the present and the insects which are now found on such trees very largely represent the natural or very early insect fauna," Marlatt wrote.[9]

The couple spent seven months touring the country, often by rickshaw, but Marlatt found no evidence that San Jose scale came from Japan. He did, however, discover one clue that the disease had spread to Japan from China.

In October 1901 the Marlatts arrived in Chefoo, China, and met representatives of the Inland China Mission. They told him about the late Rev. John Livingston Nevius, a Presbyterian from upstate New York who had spent forty years in China. Nevius had established an orchard on the mission's grounds and often exchanged cuttings with growers on the west coast of the United States. Marlatt examined Nevius's trees and saw that they were slightly infected with San Jose scale, a sign that suggested a connection to James Link.

Marlatt found the next and most important clue when he ventured deeper into China. The couple went to Peking, where they found a market selling apples, pears, and persimmons grown in a remote area nearby. This fruit was also slightly infected with the scale, indicating that although the disease was present, its natural enemies had controlled its spread.

Marlatt learned that the fruit in Peking markets came from a corner of the country cut off from the rest of the world on all sides by the Great Wall, the eastern Gobi desert, and Manchuria. He concluded that these barriers had contained the pest, perhaps for centuries, until a slight crack opened in China's relations with the world. That set off an inevitable botanical chain of events with potentially devastating consequences.

The first link in the chain was a western missionary, probably Nevius, who bought fruit grown behind the Great Wall. That fruit infected his trees, but the pest's natural enemy protected them. Nevius sent at least one tree to James Lick in California, where no natural protection existed, and it infected his whole orchard. When one of Lick's trees was shipped to the East Coast, the scale from behind the Great Wall threatened to destroy millions of fruit trees throughout America. "It is the writer's belief that [Lick] imported from China, possibly though Dr. Nevius or some other, the flowering Chinese peach and brought with it the San Jose scale to his premises," Marlatt wrote in 1902.[10] He tried to correct the record by changing the pest's name from the San Jose scale to the Chinese scale, but he failed.

After Marlatt found the disease's source in China, he had no trouble identifying its natural enemy, which was another tiny ladybird beetle. He collected all he could find in China and Japan and shipped 175 to America, many in tiny cages. About 30 survived the voyage to Washington; only two lived through the first winter. One survivor was pregnant, however. She produced more than 5,000 ladybird beetles and created a dynasty that has helped keep the San Jose scale under control in America.[11]

After he finished his work, Charles and Florence Marlatt spent most of the rest of their journey sightseeing. They visited Java and toured many places Fairchild had visited when he was studying there. Like everyone else, Marlatt commented on the magnificent view of Mount Salak from his room in the Hotel Bellevue in Buitenzorg. While they were in Java, the couple socialized with Melchior Treub, the botanical garden's director. They learned that Treub was still angry about David Fairchild's sudden departure five years earlier. "Before he had more than started on what was to have been a long period of studies which had been mapped out, [Fairchild] allowed himself to be picked up and taken away by his globe-trotting friend," Treub complained to Marlatt, who included the anecdote in a book he wrote more than fifty years later.[12]

Only one month after this conversation, Marlatt was shocked to run into Fairchild and Lathrop in Singapore as they were boarding the SS *King Albert* for Ceylon. "At the top of the ladder one of those one-in-ten-thousand chances happened—David Fairchild was standing there, still globe-trotting and plant collecting with Barbour Lathrop," Marlatt recalled. The four Americans were glad to see each other. They enjoyed eating, talking, and even botanizing together on the voyage, although within days Florence Marlatt got sick. She was "slightly invalided by some microbe she had acquired in the last days in Java," Marlatt explained at the time.[13]

After the group arrived in Ceylon, Florence Marlatt stayed in bed for several weeks. She still hadn't recovered fully when the Marlatts returned to Washington in the spring. "Mrs. Marlatt is very much better," Marlatt told Fairchild on May 20, 1902. "She is not entirely recovered, I regret to say, but I hope that she will be out of all her difficulties very soon."

Marlatt didn't mention that his wife was a devout Christian Scientist who refused medical treatment and depended entirely on prayer whenever she was sick. She did recover from the tropical infection in 1902, but a year later she became much sicker. Because she believed that prayer had cured her the first time, she again refused to see a doctor. The second time prayer failed.

On October 29, 1903, Florence Marlatt, thirty-eight years old, died from a stomach ailment at the couple's home in Washington. She "was strong in her faith to the end," reported the *Washington Post*, although her husband had pleaded with her to let him call a doctor for help.[14]

In 1905, when he served as best man at David Fairchild's wedding, Charles Marlatt was an attractive, eligible widower. Marian Fairchild wasted little time helping him find a second wife. She introduced Marlatt to Helen Mackay-Smith, a childhood friend who was the daughter of a wealthy Episcopal minister who had been the rector of St. John's Church near the White House. The couple married fourteen months later. David Fairchild was, naturally, the best man at their wedding.

*   *   *

After Marlatt solved the mystery of San Jose scale and traced its spread from China to Virginia, he continued to insist that limiting plant imports would be pointless. He even wrote in the 1902 *Yearbook of Agriculture* that the government had been too tough in its drive to control San Jose scale. "Its importance has undoubtedly been exaggerated and the restrictions imposed in consequence on the interchange of fruit and vegetable products are unnecessarily severe," he argued,[15] expressing an opinion that Fairchild shared.

The two men remained close friends. Soon after the Fairchilds were married, they moved in with Marian's parents while they were building a house in Maryland. When Mabel Bell wanted to know how to make Fairchild feel at home, she asked Marlatt for advice. Marlatt knew him well enough to recommend "plenty of writing paper and a simple diet without pies and cakes," Mabel Bell said in May 1906. Fairchild taught his children to call Marlatt "Uncle Charles," just as they were told to call Lathrop "Uncle Barbour."[16]

One strong bond between the two men was their appreciation of flowering cherry trees, the rare, glorious plants they had each seen in Japan. Marlatt planted a double-flowered variety in his yard on 15th Street in Washington, and every spring he hosted a Japanese tea party to celebrate the tree's blossoming, just like the emperor of Japan.

*   *   *

Leland Howard did not, however, agree with Marlatt and Fairchild. He continued to ask Congress for the authority to block foreign plant imports. No bill went anywhere, although Howard managed to persuade the agriculture department to change its internal rules.

In 1906, the agency ordered entomologists to examine all seeds and plants received from Fairchild's explorers. If inspectors found any pests, the new regulations said, they were ordered to fumigate the plants and, if that process didn't kill the insects, the entomologists had to destroy the material.

Fairchild initially supported the change because he didn't want to import any dangerous foreign pests or diseases. Since he had written his bulletin on systematic plant introduction in 1898, the number of serious problems had increased. The biggest threat came from the gypsy moth, which had begun devouring leaves of hardwood trees in Massachusetts in 1889 and quickly spread throughout New England.

On January 11, 1907, Marlatt asked the House Committee on Agriculture for extra funds to help Massachusetts fight the gypsy moth. He was candid when he told committee members that an extermination campaign would be expensive and probably futile. "These problems are big problems; they are not stopping," he testified. "It is like disease: we have to have doctors all the time. Even if they cure us of one attack we get another."

The gypsy moth infestation preoccupied Howard at this time, so he assigned Marlatt, whom he said had always been "clear-headed, able and versatile to a degree,"[17] responsibility for the fight for a law restricting plant imports. The first step was to persuade lawmakers to give federal entomologists the power to register, inspect and, if necessary, quarantine all imported plants and trees, not just the material brought in by the department's foreign plant explorers.

Although accepting the assignment violated the beliefs he had expressed for years, Marlatt, a dutiful subordinate, moved fast. He drafted a version of the legislation, which the House approved at the end of January 1909. The Senate got the bill in the middle of February. Within two days the Senate agriculture committee voted to send the bill to the full Senate for approval, marking more progress in a few weeks than Howard had seen in more than twelve years.

In his haste to push the bill through, however, Marlatt had failed to comply with the political rules of Washington, D.C. He never contacted commercial nurseries, the industry that would be most affected by the dramatic policy change. Quarantining fruit trees from France, for example, could have ruined many American businesses. As the legislation was speeding toward final passage, William Pitkin, the nurseries' top lobbyist in Washington, realized what was going on and moved faster than Marlatt.

To delay a final vote, Pitkin asked Marlatt to recall the bill for minor changes. Marlatt, who had never before been responsible for shepherding a bill through Congress, agreed. Because of this tactical error, the legislation to restrict foreign plant imports languished in committee for three more years. In the meantime, Fairchild's plant explorers kept hunting.

proof

Plant Enemies

After the lobbyist outmaneuvered him on the plant quarantine bill in 1909, Charles Marlatt stopped thinking like a scientist and started acting like a politician. He was, in Howard's words, "a man who did not allow himself to be discouraged."[1] Marlatt was a decisive, often arrogant man whose ambition drove him to do whatever was necessary to get what he wanted. Determined to rise from his obscure position as a second-ranked scientist in a federal agency, Marlatt waged a tough campaign to ban foreign plants. Ironically, a clash with his old friend David Fairchild helped give Marlatt the victory he wanted.

Despite their long friendship, friction between the two men was inevitable. Fairchild was a romantic idealist who believed passionately in the free exchange of plants among people around the globe. In the years between the Spanish-American War and the First World War as America was setting out on the long road to globalization, Fairchild eagerly sought food and plants from other lands, the more exotic, the better. His goal was to incorporate them into the American landscape and breed them with domestic crops, to produce offspring that were superior to existing plants.

But Marlatt was different. Instead of embracing the cosmopolitan spirit sweeping the country, he fought to protect America from the debilitating infections that foreign insects could transmit. As an entomologist, not a plant breeder, Marlatt wasn't concerned with the benefits of combining domestic stock with foreign material. While Fairchild called his imports "plant immigrants," Marlatt called them "plant enemies."[2]

This growing suspicion of foreigners—botanical and, thanks to the efforts of Davenport, Laughlin, and Paul Popenoe, human as well—soon made Fairchild's mission harder than ever to accomplish.

* * *

Xenophobia had always existed in America, but organized drives to keep out foreigners solidified only at the end of the nineteenth century. Antipathy existed against many ethnic groups, of course, but at the start of the twentieth century, prejudice was strongest against Asians.

California had limited the right of Chinese workers to immigrate since 1884, and by the turn of the century, discrimination on the West Coast had spread to other nationalities. In 1906 the San Francisco school board ordered that Japanese and Korean children be segregated with Chinese in Asian-only schools. That school board's vote was the first official indication that not everyone welcomed the Japanese people who were sailing to America in increasing numbers.

To deal with the political problems of this growing resentment against Japanese immigrants, President Theodore Roosevelt and the Japanese government reached what they called "a gentlemen's agreement" that restricted the number of Japanese workers who could legally enter the United States. In 1907, when the agreement was enacted, the number of Japanese immigrants was almost 31,000 a year; the next year it plummeted to about 16,000.

As America's suspicion of Japan grew, so did David Fairchild's affection for sakura, the country's sacred flowering cherry trees. To the Japanese people, the trees' blossoms have a mystical importance. Twelve centuries before Fairchild learned about the trees, the emperor and his court established the custom of picnicking beneath them each spring during the blossoms' brief period of glory, a flowering that traditionally symbolized the grandeur and transience of life.

The trees were virtually unknown in America in 1903, when Barbour Lathrop paid for the special shipment of thirty varieties that died in California. By the time David Fairchild married Marian Bell, he was determined to try to introduce them again. He even found a spray of blossoms for her to carry at their wedding, but in her excitement she forgot to take it to the ceremony.

Early in 1906 the couple ordered 125 cherry trees—at 10 cents each—from the Yokohama Nursery Company. Despite the common belief that the trees could not grow in Maryland, the Fairchilds had little trouble establishing them on a slope on their property, which they had named "In the Woods." "It was all an experiment in those days," Fairchild explained. "Doubts of their hardiness had been expressed by so many horticultural experts that I tried to coddle them by planting them in sheltered spots."[3] Fortunately, the Fairchilds had hired a Japanese gardener who helped the trees thrive and deepened their appreciation of them. He taught them the English translations of the trees' names: "The Tiger's Tail," "The Milky Way," and "The Royal Carriage Turns Again to Look and See" were varieties in their orchard.

David Fairchild loved these cherry trees, perhaps as much as he loved any plant. When the trees were big enough to flower, he regularly wandered through his private grove to try to capture their beauty in photographs. "I used to roam among our trees at dawn and gaze at the individual flowers through the darkness of my enlarging camera before the dewdrops had vanished from their petals," he wrote. "I seemed a Lilliputian wandering among their soft, velvety surfaces."[4]

In 1908, after the Fairchilds and their gardener proved that the trees could flourish in the area, the couple ordered trees as gifts for all the schools in Washington. When the shipment arrived, each school sent a boy to the Fairchilds' home to pick up its tree and get a quick lesson on how to care for it.

As anti-Japanese sentiment spread from California across the United States, Fairchild grew determined, he explained later, to show the American people "how deeply there lies in the Japanese character a love for the beautiful."[5]

During an elaborate Arbor Day celebration later in 1908, Fairchild boldly predicted that one day the District of Columbia would be so famous for its flowering cherry trees that people would come from all over the country to feast on their beauty.[6]

Fairchild wasn't the only person who wanted to bring the trees to America. Eliza Ruhamah Scidmore, an Iowa-born writer who often visited Tokyo and wrote popular travel articles about Asia, had been trying for more than twenty years to persuade officials in Washington to plant cherry trees

FIGURE 18. Eliza Scidmore as a young woman. By permission of the Wisconsin Historical Society, image no. WHS-90045.

around the city. She was the first female board member of the National Geographic Society and a longtime friend of Alec Bell and his family. In 1908 she joined forces with Fairchild and, soon afterwards, the new president of the United States.

In March 1909 William Howard Taft took office. He had lived in Asia as governor of the Philippines and learned to appreciate Japanese flowering cherry blossoms. As soon as Taft and his wife, Nellie, moved into the White House, Scidmore told the First Lady about her plan to beautify the city with imported cherry trees. The city's new Speedway had recently opened in Potomac Park, creating what Scidmore, Fairchild, and, in short order, Nellie Taft thought would be the perfect spot for a cherry orchard.

After Nellie Taft nudged American and Japanese diplomats to support the plan, things started to happen. In the fall of 1909 a rich Japanese American chemist bought 2,000 trees and arranged for the city of Tokyo to give them to Washington as a symbol of international friendship. Scidmore called it "an admirable Samurai retort" to America's growing prejudice against Japan.[7] In December 1909 three railroad cars filled with cherry trees arrived in Seattle, where officials of the agriculture department made a preliminary inspection and found them to be free of pests.[8] The trees traveled by train across America and arrived in Washington on January 6, 1910.

Then Charles Marlatt started causing trouble.

* * *

After the trees were unloaded at the government greenhouses on the Mall, Marlatt had a chance to promote his proposed law. Ignoring the clean bill of health granted in Seattle, he demanded a second inspection to check for foreign plant pests and diseases. For one full week, while Fairchild fretted, six experts from the agriculture department carefully examined each tree to see if it was healthy. Fairchild couldn't sleep during the nights of uncertainty. "I had been worried about the trees, fearing that they might prove too large, but I had not dreamed of any difficulty with the quarantine authorities," he recalled later.[9]

Much more was at stake than the fate of the trees. Marlatt's bill undermined Fairchild's lifelong commitment to the free exchange of plants and

FIGURE 19. The first shipment of flowering cherry trees from Japan was destroyed on January 28, 1910. Photo from the David Fairchild collection, Fairchild Tropical Botanic Garden, 2775.

ideas among nations. It also pitted politics against science, often a dangerous game in Washington, D.C.

Marlatt's team finally announced its findings on January 19. They had discovered a cornucopia of bugs and diseases: root and crown galls, scale insects, worms, weevils, shiny black ants, and, worst of all, *Lepidopterous larva*, tiny hidden worms that bore holes inside trees. The debacle was, in Fairchild's words, "a hornet's nest of protesting pathologists and entomologists."[10]

Marlatt immediately informed Secretary James Wilson that all the trees must be destroyed immediately because of the possible but unseen threat from the woodborers. "Twenty percent of the trees are visibly infested with this insect," Marlatt told Wilson in a letter dated January 19, 1910, "but it is impossible to tell how many of the others are also infested since discovery is only possible in the later stages when the insect has burrowed to the surface." He decreed that fumigation, which would have preserved the trees, would not eliminate the danger. "The entire shipment should be destroyed by burning as soon as possible," he wrote.

Marlatt's findings were so technical that no one argued against them, although a *New York Times* editorial writer was skeptical: "We have been importing ornamental plants from Japan for years and by the shipload, and it is remarkable that this particular invoice should have contained any new infections."[11]

On January 28, after three weeks of anxiety, President Taft issued the formal order to destroy the trees lest they contaminate domestic plants. Charles Henlock, the White House's head gardener, gathered the trees and the bamboo poles and packaging they arrived in, stacked them against each other like Indian tepees, and set them on fire. They were dry from their long travels and ignited quickly as news photographers recorded the destruction. Tokyo's gift of international friendship was soon reduced to ashes.

News of the conflagration was published on the front page of the *New York Times* and triggered a two-page article in the *Washington Post* about the dangers of foreign plants. The Washington *Evening Star* reported that America may have "escaped an insect yellow peril."[12] Finally, Marlatt's bill had the country's attention.

* * *

On April 27, 1910, while the sensational news of the giant bonfire still reverberated throughout Washington, Marlatt and Howard returned to Capitol Hill to pressure the House of Representatives to ban foreign plants. This time Marlatt testified that he agreed with Howard. The two entomologists emphasized the high cost of controlling plant pests and asserted that more than half the pests in America had come from overseas, an invasion they estimated cost Americans about $400 million a year (almost $10 billion in current dollars.) "The United States occupies a unique position among the first-class nations of the world in that it has no national legislation to prevent the introduction of plant diseases and of insect pests," Howard said.

They pushed the committee especially hard to pass the bill's most significant provision: granting the U.S. Department of Agriculture the authority to block any imports, including huge shipments of apple and other fruit trees from France, the biggest single supplier to the America's commercial nurseries.

Marlatt's comments were dramatic. "Plant diseases are like human diseases," he testified. "They are in the blood, in the plant itself, and it is much safer to shut off importation, because it is much better that there should be a loss of two thousand dollars to a nurseryman than a loss of ten million dollars to the country at large." He said that barring all nursery stock would be the best solution to the problem, but he didn't ask Congress to go that far yet. "I gave it as my personal belief that the country would be safer if nursery stock were excluded," he said, "but the idea of the bill is not to exclude it."

Gone was the scientist who reasoned that quarantines were ineffective and stated that it was futile to try to control nature. Marlatt said nothing when Representative William Rucker, a Democrat from Missouri, questioned Howard about the efficiency of an inspection system.

"Is it not the easiest thing in the world, doctor, for one of these little things, an eighth of an inch or less, to get away any time and escape inspection?" Rucker asked.

"Not if the states and the department are working in perfect harmony," Howard answered.

Echoing the xenophobes who were trying at the time to keep human immigrants out of the country, Marlatt discussed the importance of freeing America from foreign influence. "The nurserymen of this country, many of them, grow their own [root] stock. The importation of such stock is therefore to some extent in competition with the home business," he said. "It has developed that certain stock can be grown cheaper in France and Holland than here, and I believe the nurserymen claim that the stock for grafting purposes is better foreign grown than home grown. That may be a sound argument, but there can be no doubt, if this country should set up a Chinese wall against such importations, we could take care of our own needs."

The nursery owners feared most what Marlatt would do with the power to control all plant imports. William Pitkin, a nursery owner from Rochester, New York, who was still the industry's top lobbyist, said commercial growers were "afraid to put ourselves into a noose which might be drawn tight someday and especially if one end of that noose is going to be held by our friend Mr. Marlatt," he told the committee. "We want to keep out of it just as long as we can."

Pitkin helped kill the bill that session, too, but Marlatt refused to give up the fight. One way he kept the issue alive was by challenging Fairchild's cosmopolitan position in public debates. In April 1911, the *National*

Geographic Magazine, always open to encouraging discussion of scientific policy, published an article by Marlatt about the dangers and costs of imported plant pests. It was illustrated, in the magazine's tradition, with stark photographs of trees stripped bare by gypsy moths and close-up shots of infected fruit and tree bark. In the article Marlatt raised his estimate of the annual cost of foreign plant pests from $400 million to more than $500 million. Marlatt focused hard on the hazards of trading with China. "Our increasing business relations with China and other Oriental countries adds enormously to the risk of the importation of new pests," he stated. Marlatt knew that virtually all plant imports from China came from Frank Meyer, Fairchild's top explorer.

Six months later, Grosvenor published a counter article by Fairchild about the benefits to America of the 31,000 plants his office had already imported. "New Plant Immigrants" detailed the value to the nation of the work of Meyer and other explorers and reemphasized Fairchild's firm belief in the importance of his work.

During this public debate Marlatt's bill remained in committee until the end of 1911 when a dramatic discovery in the Pacific pressed the matter. Inspectors in Hawaii found the pest they dreaded the most: the highly contagious Mediterranean fruit fly. That insect, which fills fruit with larva known as maggots, is virtually impossible to eradicate after it appears. Growers on the West Coast were terrified that it would spread to the continental United States and wipe out their crops. The California legislature met in special session in December 1911 for the sole purpose of approving a resolution pleading with Congress to enact the national plant quarantine law.

While officials waited for help from Washington, state inspectors scoured all cargo and personal baggage—even little old ladies' purses—to stop any piece of possibly infected fruit from entering California. The pressure on Congress was intense. At another House agriculture committee hearing in January 1912, Representative Everis Anson Hayes, a California Republican, explained the seriousness of the new threat. "The fact of the matter is in a hundred years no pest has appeared that has so disturbed the fruit industries of the world as has this one," Hayes said.

West Coast farmers' panic compelled members of Congress to pressure Marlatt and Pitkin to reach a deal, although Pitkin was still angry about

Marlatt's refusal to negotiate the original bill with him. Pitkin had admitted during the last public hearing on the legislation, held on February 19, 1912, that it was Marlatt's imperious attitude that caused the delay. "We were not consulted, we were not invited, but we butted in," he said. "I think if there had been an opportunity for such a joint conference that a good deal of friction and misunderstanding might have been avoided."

The two sides reached a deal, but Marlatt admitted that the final bill was a major compromise. "It does not represent the department's views," he said, "but it will be an immense benefit to the country. It will cover the fruit fly. It will cover any new danger that comes in."

In the summer of 1912—after fourteen years of pressure from Howard and Marlatt—Congress approved restrictions on foreign plants. President Taft signed it into law on August 20, 1912, effective immediately. International plant hunting would never be the same.

*   *   *

That March, while negotiators were writing the final version of the bill, Tokyo's mayor gave Washington a second batch of flowering cherry trees. To avoid the embarrassment of any more pests and diseases, Japanese gardeners had grown these trees especially for America and kept them segregated from other plants. This batch arrived in Washington on March 26, 1912, and immediately passed government inspection. The next day the trees were quietly planted on the north side of the Tidal Basin in West Potomac Park. Nellie Taft attended the simple ceremony, along with Viscountess Chinda, the Japanese ambassador's wife, but Marlatt was absent. This long-awaited event established the cherry trees that, as Fairchild had predicted, bring thousands of visitors to Washington each spring.

*   *   *

The debate over imported plants did not end with Marlatt's legislative victory. When the law went into effect, Secretary James Wilson named Marlatt chairman of the Federal Horticultural Board, the powerful new body created to enforce the regulations. Overnight, Charles Marlatt alone had the authority to block imports from any place he wanted. "We have found that we have a good deal more power under that act than we thought we would have when it was passed," Marlatt admitted to a House committee on December 9, 1913.

A provision of the law permitted David Fairchild's office to continue to import seeds and plants as long as they were carefully inspected, treated, and, if necessary, destroyed. Yet Fairchild claimed Congress didn't give him enough money to do that job properly. By 1914 he was scrambling to handle the influx of material that Meyer and the other explorers had risked their lives to find.

He received only $5,000 (about $110,000 today) to build a receiving station on the Washington Mall, not enough money to keep the plants healthy until they were inspected. Space was so limited that shipments were often stored on the sidewalk outside the station. Fairchild deeply resented the lack of financial support for foreign plant introduction. "There were times when I felt that my old friend Charles Marlatt and his associates would gladly have done away with the introduction of plants from abroad altogether," Fairchild wrote in his autobiography. "But I determined the door should not be shut. As long as I was able, I would keep my foot in the doorway and prevent importation from being entirely forbidden."[13]

Marlatt's staff continued to discover dangers lurking in Fairchild's imports. In 1916 the Federal Horticultural Board announced that it had found 157 diseases on plants imported by the office of seed and plant introduction, most of them from Meyer in Asia.[14] Yet there is no evidence that any serious disease or pest actually established itself in America through Fairchild's introductions.

In 1917 the American Forestry Association invited Fairchild and Marlatt to debate the issue again at its annual meeting in Washington. Marlatt went first. "The importation of new stock in the last few years from these [Asian] regions by the Department of Agriculture and by private agencies has especially demonstrated the existence therein of many very dangerous plant pests," he said, sparing his old friend no embarrassment. For the first time, furthermore, Marlatt raised the possibility that he might seek authority to keep out all foreign imports. "It is perhaps opportune now to seriously consider the advisability of very much restricting the further entry of all foreign plants and plant products," he said.[15]

When his turn came, Fairchild fought back, insisting that imported plants had already brought many benefits to America's economy and to global horticulture. And, he said, they were likely to bring even more in the future. "I cannot feel the same degree of confidence which some people seem to have," Fairchild said in an obvious reference to Marlatt, "that we can decide now a policy which will protect these little seedlings for the next hundred years in the face of the gigantic changes in transportation and commerce which those years will produce."

He said Marlatt's policy reflected a narrow, isolated, and out-of-date sense of America's role in the world. "We can say to ourselves: 'Let us be independent of foreign plant production. Let us protect our own by building a wall of quarantine regulations and keep out all the diseases which our agricultural crops are heir to and have this great advantage over the rest of the world,'" Fairchild said in this 1917 speech. "But the whole trend of the world is toward greater intercourse, more frequent exchange of commodities, less isolation, and a greater mixture of the plants and plant products over the face of the globe."[16]

*　　*　　*

Marlatt refused to back down. In 1918, he pulled the end of the noose, as Pitkin had feared, by announcing plans to bar virtually every foreign plant. The 1912 law gave him the power to do it by administrative fiat, without congressional approval.

Pitkin tried again to stop him. "There is absolutely no necessity for such a radical change in existing law," Pitkin insisted on Capitol Hill in February 1918. He took particular issue with Marlatt's claim that the United States was already spending $500 million every year battling foreign pests. "The figure might just as well been placed at five hundred billion dollars as it is entirely guess work and undoubtedly far from the real truth," Pitkin said.

Yet this time Pitkin was powerless. Marlatt gave the industry one year to prepare for the new rules, but he delivered on his threat. He blocked foreign plant imports to America on June 1, 1919, under an administrative act titled Plant Quarantine No. 37. "The only safeguard for the future was the exclusion of all plant stock not absolutely essential to the agricultural, horticultural, and forestry needs of the United States," Marlatt said in his decree.[17]

As the implications of Marlatt's order sank in, criticism intensified. The editor of *World's Work*, a popular magazine, wrote in its July 1922 issue, for example, that Marlatt's expanded quarantine was the product of "seven years of gestation in the mushroom caves of an obscure office in Washington. [It] declares that the whole world outside the United States is a horticultural pest-house and closes the door against all importations of plants from them."

Marlatt's rules applied to everyone, even the respected Arnold Arboretum in Boston, which had been importing trees and herbarium specimens for educational purposes since 1874 apparently without ever introducing a plant pest or disease.

Defending himself in another *National Geographic Magazine* article, Marlatt encouraged comparisons between unwelcomed foreign plants and humans. "These undesirable immigrants we must lodge and board forever, but we want to shut the door if we can to their brothers and sisters and cousins and aunts!" he wrote.[18]

Marlatt's critics persisted. Their anger peaked in 1922 when florists demanded that Henry C. Wallace, the secretary of agriculture, overrule Marlatt and admit narcissus and other flowering bulbs from Holland. Although Wallace rebuffed their demands, the arguments continued.

Stephen F. Hamblin, director of the Harvard Botanic Garden, castigated Marlatt's anti-immigrant position. "If such a policy had been enforced a hundred years ago, America would lack at the present time more than half the plants that make her gardens beautiful and more than half the fruits, grains, and other economic plants that make her horticulture profitable and advantageous," Hamblin wrote in a March 1925 *Atlantic Monthly* article.

Like many critics, Hamblin believed that the law gave Marlatt too much power. His position "appears to combine executive, judicial, administrative, and advisory functions and, in cases of appeal, to act as defendant, judge, and jury." Hamblin echoed Fairchild when he called for more—not less—cooperation among nations. "It is surely not too great a thing to ask in a civilized age that the unavoidable interchange between countries shall be productive of growth and ideas, as well as of pests and diseases," Hamblin wrote. "For, after all, Pandora's box was opened long ago."

The pressure on the department was intense. Although Marlatt kept his job as head of the Federal Horticultural Board, Wallace appointed someone to oversee his work in October 1923, a change that apparently made a difference. In June 1925 Wilson Popenoe told Fairchild that the quarantine restrictions were easing. Despite their disagreement, Marlatt and Fairchild's wives were close, and the two men remained friendly. A few months later, Marlatt confided that he might get fired. "The campaign against me is getting very hot," he told Fairchild on September 24, 1925, adding as a footnote, "Don't be alarmed. It is merely the machinations of the same old gang who have often been worsted in the past."

Marlatt held onto his job until 1933, but the controversy didn't die. In the spring of 1940, long after Marlatt and Fairchild had retired from the agriculture department, Congress held another hearing on the still-thorny issue of admitting flowering bulbs from Holland. By then, however, the agriculture department's position had changed. The scientists had won.

Henry A. Wallace, a plant breeder and business executive who was the son of the man who had upheld Marlatt's right to ban foreign plants, had become the U.S. secretary of agriculture. His testimony reflected the same position that David Fairchild had expressed many years before. "The over-all guiding policy," he said on April 25, 1940, " . . . is the policy that, after all, these matters should be decided on a scientific basis, and that science should not be expected to bend itself to attain economic ends."

Wallace told the Senate Committee on Agriculture and Forestry that the ban on flowering bulbs, to take the current example, had not after all protected America from the eelworm or nematode, a dreaded parasite that attacked wheat and other crops. Eelworms had become "so widespread and so impossible of eradication by any reasonable means," he said, that continuing the ban would be useless.

One example of the situation was a pest that was then attacking potato farms in Nassau County, New York. In 1941 scientists determined that the problem was the golden nematode, a pest from Europe that was probably carried not by an imported flowering bulb but in mud on the tires of military equipment stored on a farm in Hicksville after the First World War. The insect wiped out virtually all the farms and soon transformed Long Island from farmland to a vast suburb of New York City.

The agriculture secretary's statement was the long-awaited admission that—just as Fairchild had insisted many years earlier—Marlatt and Howard's ban hadn't worked. "The old regulations were ineffective and unnecessary," Vivian Wiser, a department historian, wrote in 1974, "since the pests they sought to exclude were already here."[19]

Nature had ignored national boundaries, the economic needs of American nurseries, and especially the dictates of the Congress of the United States. Efforts to keep foreign pests out of America had turned out to be, as Marlatt himself had warned the Association of Economic Entomologists on August 18, 1899, "vain attempts . . . to prevent the inevitable."

The Impossible

Five months before Charles Marlatt's bonfire in Washington dramatized the risks of foreign plants, Frank Meyer returned to Asia to find more.

On August 15, 1909, the day after he left New York, the *Washington Post* gave Meyer an enthusiastic, full-page send-off, calling him the agriculture department's own Christopher Columbus. In an article illustrated by a photograph of Meyer wearing heavy boots and a thick fur coat, the writer vividly described Meyer's assignment. "It is one of the most daring expeditions ever undertaken, for he will go alone into the wildest parts of the civilized globe," the writer said. "For weeks and months he will be the only white man amid hordes of untamed barbarians; he will be surrounded by burning sands, by alkaline plains, by frozen steppes, by mountain fastnesses, and all other dangers of which the mind of man can conceive."

Except for the untamed barbarians, the description was largely accurate. Mayer's assignment this time was to explore eastern Russia, Turkistan, Mongolia, Siberia, and Kazakhstan, each among the most remote, challenging places on the globe. He planned to spend two years crossing Central Asia from the Caucasus east to Siberia, a 2,000-mile trek.

Meyer began slowly. He stayed in England and France for a few weeks to meet other botanists and spent several days with his mother and other relatives in Holland. His expedition was delayed further in St. Petersburg because Russian officials would not give him the documents he needed to travel east. Meyer, who suspected that his refusal to pay bribes caused the delay, didn't back down even though the holdup cost him several good weeks of mild weather for collecting. Finally on January 1, 1910, officials gave him permission to leave.

Trying hard, as usual, to put his waiting time in St. Petersburg to good advantage, he collected a few interesting specimens before moving deeper into Asia. Meyer pleased Fairchild by sending him a surprise: coffee ground from soybeans. Throughout his life Fairchild enjoyed sneaking unfamiliar foreign foods onto the plates of his unsuspecting guests. He served Meyer's gift to his wife and mother-in-law and was delighted when the women claimed—perhaps out of politeness—that they didn't suspect it wasn't ordinary coffee.

In the summer of 1910, after four months of setbacks and bureaucratic delays, Meyer was finally moving. During this trip he avoided the comfortable hospitality of western missionaries and followed the dusty, mysterious trail of a Silk Road through Central Asia, one well worn by traders for centuries.

He carried a collection of watches as gifts to local leaders; the more important the leader, the better the watch. As he wandered through Samarkand, an ancient city that is now in Uzbekistan, Meyer had no companion after his interpreter quit because he couldn't stand to walk anymore. Despite the loneliness and irritating delays, Meyer seemed content. Remembering too well Washington's summer heat, he told Fairchild, who was working in his office in the agriculture department, "I would rather die than change places."[1]

Meyer's most important destination on this expedition was Chinese Turkistan, a region with some of the world's harshest terrain. Now called Xinjiang, it lies on China's northwest frontier. To get there Meyer hired a guide and a caravan of horses and drivers to pull carts of equipment through sparsely populated oasis towns and stony stretches of desert. Because officials had delayed his departure from St. Petersburg, he arrived in Chinese Turkistan in winter when the weather was severe. During one stretch, Meyer didn't see the sun for eight days, and the hot tea in his cup froze before he could drink it. Nonetheless he managed to collect fruit tree cuttings and buy seeds in local markets.

After he explored the lowlands, Meyer traveled to the Tianshan, the glorious range that divides Chinese Turkistan from Mongolia and that

FIGURE 20. Frank Meyer's caravan stops for a break in Chinese Turkistan on October 17, 1910. Photo by Frank Meyer from the Arnold Arboretum Horticultural Library of Harvard University, © President and Fellows of Harvard College. Arnold Arboretum Archives.

the Chinese called "the heavenly mountains." The area was so forbidding and mysterious that even eighteen years after Meyer's visit explorers complained they could not find accurate maps to guide them. Meyer got through by skirting the Gobi Desert.

In early March he crossed a treacherous glacier at Muzart Pass that cuts through the center of the mountain range. Meyer and his caravan of seven packhorses took six hours to slog carefully up the icy steps of the moving glacier, which lies at an altitude of 9,700 feet. "The cracking and moaning sounds heard at intervals made one feel anxious to get on terra firma," he admitted.[2]

His small group climbed beyond the glacier to 13,000 feet where the temperature was so cold that Meyer's shoes froze to his feet. "These inhospitable, cold, sterile, wind-swept mountains," he reported later.[3]

Fortunately, his ordeal was rewarded. Spring had arrived on the other side of the mountains, an area of great plant diversity where Meyer discovered hardy apples, wild apricots, and a new asparagus for American farmers.

A few days later Meyer was in a region of Mongolia reputed to be thick with murderous robbers. His guide, who was too terrified to remain with the group, fled, taking Meyer's credentials and official papers with him. Meyer and his remaining assistants were so wary of attack that, despite the bitter cold, they lit only a tiny fire to heat their dinner and extinguished it before dark. They waited without moving until dawn.

The next day, as the group reached the highest mountain peak in the eastern section of the Tianshan, Meyer found a friend, not a foe. The local chief, surprised and delighted to have visitors, invited the men to dinner in his tent. A gracious host, he killed a sheep in Meyer's honor and served the generous feast by candlelight on a table adorned with Tibetan silver and a Russian samovar. Meyer learned, however, not all visitors were welcomed the same way. "The Kirghiz guides had told us that men are sometimes butchered like sheep for sacrifice," Meyer wrote.[4] In gratitude for the fine meal, he gave his host his best wristwatch.

Next Meyer tackled the Altay Mountains, a range connected to the Tianshan that runs through Siberia. It was rich land for a plant explorer, but it provided few comforts for a traveler on foot. For days Meyer ate nothing but tea, bread, and dried sausages. The only fuel he found to boil water was dried cattle dung.

A few days later, he and his men met peasants walking to the local market. They were friendly and curious about Meyer and America, so he showed them photographs in a magazine—probably the *National Geographic*—that he carried with him. One featured a twenty-story building, a sight that amazed the peasants. Meyer reported that they wanted to know if strong winds blew down the buildings as they sometimes blew down tall trees.[5]

When spring finally arrived in the region, Meyer hiked through mountain meadows filled with gloriously colored wild flowers. In June 1911 he reached Berel near the border with Mongolia. There, amid bare but beautiful forests, Meyer met a farmer who had gotten rich selling stag antlers to men who came from distant lands to purchase what they believed was the secret of youth and virility. Meyer found these adventures satisfying. "I think I was born for the work I am engaged in now," he told Fairchild that spring.[6]

Despite many hardships, Meyer managed to collect, package, and ship several hundred new plants to America. He was proud of himself. "This agricultural explorer work is a mighty great education to the man who is able to stand it," he told Fairchild. "Pushing on . . . to conquer new worlds is the great spring that drives us on."[7]

Chinese Turkistan had been an enormous challenge—Fairchild said later that it was the most uncomfortable trip ever made by any of his explorers—yet Meyer's optimism was still strong. "I personally have very little bad luck on my journeys," he reported. "It seems as if the good wishes from so many people who take an interest in my work keep some sort of protective atmosphere around me."[8]

His luck continued through the rest of the expedition. In April 1912 Meyer had planned to sail from England to America on the RMS *Titanic*. He fell ill in London, however, and changed his itinerary. He left on the RMS *Mauretania* instead, averting almost certain death in the North Atlantic. After Meyer returned, the *Los Angeles Times* published another article praising his work. Meyer, the story said, "had a faculty of doing the impossible, getting along in impossible places and with impossible people as though that had been his chief recreation in life."[9]

* * *

While Frank Meyer was in Washington preparing for his third trip, he was heartbroken to learn from relatives in Holland that his mother was dead. He had seen her last in the fall of 1909 on his way to St. Petersburg. The loss hit him hard, much harder than other people realized. Her death was "a great emptiness that will never be filled," he told friends. "My own life has less value to me because there is no hope ever to see her again."[10]

Yet Meyer pushed on. He conscientiously prepared for his third trip by traveling across America to study local growing conditions. He visited an experiment station in North Dakota, the Arnold Arboretum in Boston, and many farms and gardens in between. The main purpose of Meyer's third expedition was to explore Kansu, a little-known area near Tibet reputed to be rich in plants and trees. Only two western naturalists, William

FIGURE 21. Seven boxes of seeds are ready to be transported from Peking to California on October 16, 1913. Photo by Frank Meyer from the Arnold Arboretum Horticultural Library of Harvard University, © President and Fellows of Harvard College. Arnold Arboretum Archives.

Purdom of England and Grigori Nikolaevich Potanin of Russia, had traveled deep enough into China to visit the province.

Meyer left New York on November 1, 1912, and arrived in Russia about two months later. But his progress after that was so slow—due to his own illness, political unrest around him, and endless logistical challenges—that he began to despair. His worries were increasing. Plant exploring in Asia was lonely, challenging, frustrating, and, Meyer was beginning to think, possibly futile.

He arrived in Peking on March 14, 1913, and explored the region around the city while he waited for permission to move on. He collected wild grapes, peaches, hazelnuts, asparagus, and other seeds. It was mid-December 1913 before Chinese officials cleared him to travel to Kansu. Conditions were so rough at that time of year that it took Meyer another ten months to get there. "There are several dark clouds hanging over these proposed journeys," he told Fairchild on June 4, 1914. "The worst is the brigandage out in northwest China. Powerful bands of these rascals are still operating in Kansu especially and at times I am considerably worried about the accounts one finds in the papers. I would hate to lose my whole outfit in an encounter with these scoundrels."

Meyer bought a rifle and a second revolver before he left for Kansu. "Of course we carried our firearms all the time, but luckily we had no encounter," he told Fairchild on August 1, 1914. "All we saw was a human head hanging in a little wooden cage hanging in a wild apricot tree along the roadside and grinning at us with its white teeth, showing partly through the dried-up blackened skin."

When Meyer finally crossed the border into Kansu province the next month, one of his first stops was a village where the White Wolves, the most brutal of the local gangs, had massacred all inhabitants a few months earlier. Meyer, fortunately, did not encounter the dreaded thugs.

On October 10, 1914, he reached the town called Chiehchou and wryly described his lodgings to Fairchild. "Imagine an overcrowded inn, with merchants and coolies shouting and having angry disputes; with partitions between the rooms so thin as to make them almost transparent; with people gambling with dice and cards all night long; others smoking opium; hawkers coming in, selling all possible sorts of things from raw carrots to straw braid hats; and odors hanging about to make angels even procure handkerchief," Meyer wrote. "Here you have a picture of the best inn in town."[11]

The accommodations were not the worst part of the Kansu trip. On October 23, 1914, as Meyer and his few assistants were walking on the road toward a tiny village called Siku, they were surprised to encounter two western plant explorers leading another party. The men were Reginald Farrer and William Purdom, Englishmen on a private expedition financed by Farrer to find alpine trees. Meyer had heard that Purdom, one of the two westerners who had already visited Kansu, was nearby, so he handled the unexpected meeting gracefully. Farrer, however, did not. "The coincidence took our breath away," he wrote later. "Such a thing to happen in so remote a tiny point in the world as Siku where not two foreigners have appeared before from age's end to age's end!"[12]

Plant explorers were a competitive bunch, so the westerners didn't get along well, especially after Meyer was forced to ask the others for help. He had again given himself such a challenging, dangerous assignment that local people wouldn't work for him. Before he left for Kansu he had found an interpreter for the expedition only by placing a help wanted ad in a Peking newspaper, an unprecedented hiring practice for an international plant explorer.

The inexperienced interpreter was wary of the work from the beginning and grew increasingly apprehensive as the team neared its destination. Finally at Siku he was absolutely terrified of being attacked by Tibetan bandits and refused to continue. "I will stay here and take home your body when it is recovered," the interpreter told Meyer. Meyer, furious, fired him on the spot and insisted he return the salary he had advanced. The interpreter refused. Meyer got so angry that he shoved him down some steps.[13]

Meyer didn't report the dispute, but Farrer did. With a nasty tone, he recounted the incident in an article published by the *Gardeners Chronicle* in London. "Mr. Meyer had been ill advised enough to bring with him up-country a very expensive fine gentleman of an 'interpreter' from the coast, of lily hands and liver to match," Farrer wrote in the July 3, 1915, edition. "This dignitary (who was considered in the town the Grand Seigneur of the party) once arrived at so remote and barbarous end of the world as this, flatly refused to accompany his employers any further through the danger-lands when we had just returned without trouble or hint of trouble."

Meyer always hired local interpreters because he didn't speak Mandarin or any local language, although he could manage a few phrases. The Siku villagers sided with the interpreter and scorned Meyer for his handling of the dispute. Farrer supported the Chinese. "In the eyes of Siku, [firing the interpreter] took on the aspect of an outrage," Farrer wrote in another account. "And in this case it was the eyes of Siku that counted, for now Mr. Meyer, left servant-less and solitary, was quite unable to procure any service for love or money from anybody in the place."[14]

Farrer and Purdom, who each spoke the local dialect, helped Meyer find another interpreter before they left on their own expedition, but he didn't last long either because, according to Meyer, he was an opium addict.

The entire incident provided Farrer with a juicy but probably unjust anecdote for a book about his travels that he published in England. "Meyer maintained so rooted a repulsion for everything Chinese that he successfully avoided any acquaintance even with the language . . . and therefore had an interpreter," Farrer wrote.[15]

Meyer responded to this attack a year later by pointing out that twenty-two languages and four hundred dialects were spoken in China, and it would be impossible to learn each one.[16] He didn't mention another important fact: his assignment was to travel throughout China while Farrer and the other plant explorers supported by big companies or private fortunes concentrated in one area with one dialect and one team of local employees.

Despite the lack of support from his western colleagues, Meyer continued onto Kansu and gathered material for the agriculture department and the Arnold Arboretum. He collected at least two hundred varieties of fruit and flowers—including a peach he named for Potanin, the Russian naturalist—from wild areas, temple gardens, and Christian missions. Meyer was pleased that he finally accomplished his goal, despite the hardships.

"At last I have arrived here in the provincial capital of Kansu," he wrote to Fairchild on December 10, 1914, from Lanchowfu. "I feel like an old-time sailing ship that has come into port, loaded full of all sorts of things. But the ship has weathered some storms and it is with the loss of the main sail that it is berthed here now."

Getting out of the province was more dangerous than getting in. While they were crossing the Kansu-Shensi border in December 1914, Meyer and his Dutch assistant got into a fistfight with Chinese soldiers who suspected they were smuggling opium among their seeds and plants. The soldiers' suspicions were increased by the absence of an interpreter to explain Meyer's unusual work and by Meyer's insistence on walking instead of being carried in a sedan chair like other westerners. The inspector grilled him about his plants: "Have you got any opium with you? Have you got any poppy heads? Have you got any poppy seed?"

At one dramatic moment during this interrogation, the guards pushed Meyer and his assistant up against a wall, apparently preparing to execute them, but for some unknown reason they were spared at the last minute. The ordeal ended with one guard showing his disgust toward Meyer. He "spat me in the face," he told Fairchild.

*   *   *

A few months later, Meyer's circumstances declined even more. At the end of 1914 he carefully packed hundreds of seed packets he had collected over eighteen months in Kansu and other areas and shipped them to the United States. In August 1915, while this material was being temporarily stored in Galveston, Texas, in transit between China and Washington, D.C., a hurricane hit the city. The storm wiped out the waterfront warehouse that housed Meyer's collection, destroying almost everything. The loss highlighted the increasing difficulty of landing shipments in Washington ready to be tested at experiment stations. That challenge increased as Charles Marlatt gained more power to restrict all plant imports.

During his third expedition, another hazard of Marlatt's policy was evident after Meyer sent a large shipment of peach and other fruit trees to America by way of Japan. In Kobe inspectors severely damaged many of these plants and told Meyer he could no longer relay his shipments through Japan, a country whose exports were hit by Marlatt's crackdown. The ruling created a major logistical problem for Meyer.

He was furious about the new quarantine rules—he called Marlatt's policy "throwing out the baby with the wash water"[17]—that threatened to render the hard work of collecting plants absolutely fruitless. And he wasn't grateful when Marlatt named a new insidious insect found on a plant from China *Aspidictus meyeri*.[18]

The fights in Washington, world politics, the living conditions in Asia, and his deep loneliness had begun to wear Frank Meyer down. In June 1915, after the First World War began in Europe, he went to the movies in Shanghai to try to relax, but the evening did not cheer him up because the entertainment included a newsreel showing fighting in the Vosges region of France. "It was sickening to see how many figures dropped out of the ranks," he told Fairchild. "Some were quiet straight away; others moved their arms or legs or turned over a few times and the others moved on, just as if it was the most common, everyday affair."[19]

When he first started exploring, Meyer had vowed that after he turned forty he would settle somewhere in America and spend the rest of his life contentedly testing and breeding plants. As that birthday neared, however, his plan seemed doomed. Yet he couldn't resign from the department because pensions or other retirement benefits didn't exist for civilian government workers. Instead, he dropped hints in his letters about switching to another position in the department, but Fairchild never followed up on them or offered him an easier job.

Shortly after his third expedition ended in October 1915, Meyer turned forty without any change in his situation. Instead of planning a comfortable retirement, he faced endless days of walking through Asia and struggling to deliver his plants safely at home.

* * *

In the summer of 1916, Fairchild pushed Meyer to make a fourth trip to Asia that would keep him there through 1919. As he prepared to leave America this time, Meyer admitted that despair had replaced his upbeat attitude about his "beautiful job." "The specter of a lonely old age looms larger and larger, and the spectacular office of an agricultural explorer does not hold it down any longer," Meyer wrote to Howard Dorsett, who had become his close friend, before he left Washington in August 1916.[20]

Meyer stopped first in Minneapolis to visit an experiment station near the city. Even in that comfortable region in America he was apprehensive about his future. "The long and lonely journey has commenced, and I feel the weight of it," he told Dorsett. "Is there rest and permanency somewhere?"[21]

Before he left Minneapolis, Meyer came down with malaria and was too sick with a high fever to do any work. "For a few days I was in that strange borderland when sanity has slipped and insanity is entering," he told Fairchild.[22]

While he recovered, Meyer put into storage his twenty-nine bulky bags, which weighed nine hundred pounds. "In heaven I think a fellow travels without luggage," he complained to his family.[23]

At the time the U.S. government did not pay employees when they were sick, so auditors refused to reimburse Meyer's $10.40 storage bill because he wasn't on the job. Fairchild, finally realizing that Meyer was disturbed, told him to be careful. "You are too valuable a man for the government to take chances with your health," he telegrammed Meyer before he left America.[24]

Despite his illness and anguish, Meyer kept moving west. "Life is such a short dream," he wrote his family. "People who travel realize the dreamlike quality of existence more than those who stay at home."[25]

By the time he arrived in Yokohama, Meyer said he had become resigned to his situation. "The more fatalistic one becomes, the easier things pass off," he wrote.[26]

When he arrived in China in January 1917, the temperature was low enough to freeze the Yellow Sea, slowing Meyer's progress. He was lonely and sad and in the evenings read heart-wrenching poetry by Walt Whitman. One of Meyer's favorites was "Come Up from the Fields, Father," a poem about a mother who was so heartbroken over her son's death that she wanted to die, too.

In the midnight waking, weeping, longing with one deep longing,
O that she might withdraw unnoticed—silent from life, escape and
    withdraw,
To follow, to seek, to be with her dear dead son.

In Washington, Fairchild was occupied with many other matters, especially the growing possibility of food shortages because of the war in Europe. In his letters to Meyer, Fairchild was supportive and assured him he could eventually give up plant exploring to work for the U.S. government in an easier, albeit less interesting position. He never provided any specifics, however, so Meyer saw no way out of his situation. Fairchild urged him to stay in the field. "We have only one life to live and we want to spend it enriching our own country with the plants of the world which produce good things to eat and to look at," Fairchild told Meyer on March 25, 1917.

The Fairchild pep talk didn't help, and Meyer's mood worsened. On February 3, 1917, he wrote Fairchild a long letter about his sorrow. "My own health is also not as good as I wished it to be," he said. "The loneliness of life; the great amount of work I have to do, which I can never finish; the paralyzing effects of this never-ending horrible war; and so many another thing, these often rob me of my sleep and make me feel like being a ship adrift."

In the same letter, he complained about Charles Marlatt's new quarantine rules. "I certainly hope that this idea 'kill and burn' is not going to obsess our pathologists. If so, you cannot count any longer on me staying in the service," he wrote, underlining the last phrase for emphasis.

As Meyer struggled with sickness and depression, political unrest increased in China. Finding assistants and interpreters willing to venture into the countryside became harder than ever. Meyer was also angry that the fighting in Europe made it impossible for him to travel there to see his family. On February 10, 1917, he finally admitted to Fairchild that he didn't think he could explore much longer. "The loneliness and the hardships of life here are beginning to be more and more distasteful to me," he wrote.

In March 1917 Meyer arrived in Ichang, a bustling port on the Yangtze River in Hubei province that he called "terra sancta" because it had been headquarters for both Ernest Wilson and Augustine Henry, two explorers whose work he had come to respect. "One gets a sort of feeling like a Christian who wanders thru Palestine or a Mohammedan when he sees Mecca and Medina," Meyer wrote on March 23, 1917.

Despite Meyer's miserable working conditions, Fairchild kept devising more tasks for him, weighing him down and exploiting his strong sense of responsibility. At one point, Fairchild demanded that Meyer ship one hundred pounds of opium poppy seeds to America immediately, although Chinese officials had banned the flowers and threatened to behead any farmer caught growing them. Meyer confided in Hugo De Vries, his old mentor, about his misery. "If I had seven bodies, I could use all in my work," he wrote in May 1917.[27]

Meyer was so distressed by news that the United States had entered the war that a Scottish missionary advised him to give up exploring completely unless a strict regime of exercise, cold baths, and sedatives improved his mood. Meyer, no longer the great humanitarian, grew disgusted by Chinese people he saw and started to refer to them as "human pests" and "human weeds."

At this point he sat down to take stock of his situation. He scrawled in pencil on the back of three used envelopes ten reasons why he should quit exploring. "Loneliness of life." "Squalor and dirt in China." "New plant quarantine laws." It was a sad list of conditions he was helpless to control.[28]

Meyer, who had trouble eating and sleeping, was clearly worried about himself. "I have spoken with many doctors and other residents here . . . and all say that the Yangtze River climate is mentally depressing," he wrote on May 22, 1917. In his correspondence, Fairchild was kind to Meyer, but he continued to prod him to find more plants for America. He probably didn't realize how troubled Meyer was, although he knew that it would have been difficult to replace him during the war. "You have begun a great work and it would be a tremendous pity not to carry it further, particularly during these strenuous times," Fairchild wrote on May 22, 1917.

Meyer's discoveries pleased Fairchild. At the same time that Meyer was agonizing over the sad list on his old envelopes, Fairchild served fermented soybeans—not yet called tofu in America—as an exotic appetizer to his dinner guests in Washington.

Soon, however, Meyer found himself unable to do any collecting at all because he was surrounded by fighting.

Meyer and Yao Feng Ting, his Chinese assistant, had returned to Ichang at the end of 1917 for a brief stay while Ting received routine medical care. Almost immediately, however, the city was besieged by battles between government troops and revolutionaries. Conditions were too dangerous for Meyer and Ting to leave; they were trapped in Ichang for five months. All Meyer could do was wait and watch the killing. The experience appalled him. "Last week I saw that some of these fellows took the hearts out of men whom they had shot, and mutilated the corpses in unspeakable ways," he reported on March 6, 1918. "They are going to eat these hearts to get courage!"

The same day that Meyer wrote that letter, Fairchild, always an optimist who saw the romantic side of events, wrote him a long, chatty letter full of plant talk. He even urged Meyer to collect more seeds of *Davidia involucrate*, the dove tree that inspired the exploration of China. Fairchild clearly intended his words to be encouraging. "Regardless of the fearful sorrows and the horrible features of the life around us, we must push on to bigger and grander things before life really closes in on us," he wrote from Washington.

The noble words didn't help Meyer. On May 2, 1918, the fighting finally subsided long enough for Meyer and Ting to leave Ichang and make their way, mostly on foot, to Hankow, about two hundred miles away. They stayed there for a few weeks to recover from their five-month ordeal, but the respite didn't help. Meyer was still sick and depressed, struggling to find a way to make his life worthwhile.

He sounded defeated when he wrote to Fairchild on May 18, 1918. "It often seems that we do not live ourselves any longer but that we are being lived," he said. "Uncontrollable forces seem to be at work among humanity."

Meyer was so miserable that he no longer tried to hide his situation from his family. "I start to feel tired in China with unrest and unsanitary conditions," he admitted. "The bed and walls of my room were full of lice recently, even my straw hat. I could not sleep." Even his lifelong trick of solving problems didn't work anymore. "Walk all day and feel miserable."[29]

Doubting that he could continue to collect plants because of the widespread fighting, Meyer decided to leave for Shanghai and wait for the political situation to calm down. He never got there.

Throughout his life, unconventional religious ideas had fascinated Meyer. He had studied Buddhism in Holland and was attracted to theosophy, a philosophy that seeks wisdom and divinity in nature. The theories stayed with him during his travels in Asia.

Nelson Johnson, a United States diplomat in China who became an important friend to Meyer, later recounted discussions they had had about the meaning of life. "Meyer was a believer in the transmigration of souls and professed to believe that in the previous existence he had lived as a South Sea islander," Johnson recalled years later. "I remember one night when he gave us a demonstration of a South Sea island dance which he believed he had been able to perform in that other existence of his. He blamed his mother for not having noted down the sounds which he had made as a baby as these might have given him a clue as to the particular island on which he lived."[30]

When Meyer woke in his hotel room in Hankow on the morning of May 31, 1918, he told Ting he had just had "a bad omen," a dream that his father and a few old friends had visited him. He complained of stomach trouble and a headache and said he needed to leave Hankow right away because the city was too hot.[31]

That evening the men left Hankow on the SS *Feng Yang Maru*, a Japanese passenger ship traveling down the Yangtze River to Shanghai. They were both traveling in Chinese first class because the tickets were cheaper than foreign first class.

Meyer's activities the next day were well recorded. A British insurance agent named Islay F. Drysdale boarded the ship and shared a cabin with him. The men kept each other company all day, and Meyer confided to Drysdale that the war depressed him. Later Meyer told the ship's captain he had a headache but otherwise felt fine. He ate dinner early, and when Ting brought him tea about 6 p.m., Meyer and Ting exchanged pleasantries.

He "asked me how I was getting on," Ting said later. "I asked him how he was getting on and he said he felt better than when he was at Hankow." It was their last conversation.

The river was calm that night. At about 11:20 p.m., other passengers saw Meyer walk from his cabin across the saloon toward the toilet, which was located a few steps from the ship's railing. He never returned. The next day the captain notified officials at the U.S. consulate in Nanking that Frank Meyer was missing.

Four days later, a boatman found his body floating in the Yangtze. He was wearing a white shirt, grey pants, black socks and yellow shoes. The boatman towed Meyer's body to the closest lifesaving station, where officials paid him 80 cents and gave him Meyer's shoes as a reward. His body showed no signs of violence. Local officials buried Meyer on the riverbank in the traditional way: his body was placed on two planks, and a third plank was placed on top of him. A mound of earth covered the grave.

Within a few days, a U.S. consulate official located Meyer's body and arranged for him to be buried properly at the Bubbling Well Cemetery in Shanghai. Because investigators found no suicide note, they listed "drowning" as the official cause of death. Nonetheless Meyer's colleagues said they believed he had killed himself in despair. He was forty-two, alone, and sick, his hard work diminished by politics and war. Meyer believed he had no future. "Times certainly are sad and mad and, from a scientific point of view, so utterly unnecessary," he told Fairchild in a letter that arrived in Washington three weeks after he went overboard.

Johnson wrote later that Meyer's belief in reincarnation might have made it easier for him to kill himself. "I think that he felt that he was merely leaving this life for another as grass, flower, or tree or perhaps later as another human being."[32]

When associates examined Meyer's belongings after his burial, they discovered at the bottom of his trunk the dinner jacket, green with mold, that Fairchild had given him when he started his work thirteen years earlier. Meyer had never worn it.[33]

The Last Explorers

Frank Meyer's death, combined with the political upheavals triggered by the First World War, were damaging blows to David Fairchild's work. "This loss of Meyer . . . is a very severe one," he admitted to Lathrop in a letter June 18, 1918. "I feel it personally as well as officially. . . . Meyer was a remarkable man and endeared himself to all of us through his unique personality and devotion to his work. It is now up to me to find someone else to take up the same kind of exploration work which Meyer has done and this is not going to be an easy thing by any means under the present conditions."

Meyer's death left Wilson Popenoe as America's only remaining plant hunter. If he stopped exploring, Fairchild's lifelong mission would end.

By then, twenty years after the project began in a tiny office in the U.S. Department of Agriculture, federal officials were losing interest in foreign plants. Money was tight. "The office is wailing more and more about lack of funds," Wilson Popenoe told his family on March 27, 1919.

Charles Marlatt's campaign had inflamed isolationists' antagonism against imports. The tougher regulations he enacted in 1919 included a special warning about the risks of seeds and plants from "little-known and little-explored countries," just the places that Fairchild loved most.

Even the enthusiasm of Beverly Galloway, Fairchild's mentor and champion, was cooling. Early in 1919, when the number of imports had reached about 49,000, Galloway suggested gently that perhaps Fairchild's explorers had collected enough material, at least for a while, and the office should concentrate on experimenting with the material it had already found.[1]

Although Fairchild, almost fifty years old, was also growing discouraged—Popenoe kept his parents updated about how tired he looked—he refused to give up foreign plant hunting, despite Marlatt, Galloway, and even Congress.

* * *

By the fall of 1919 Fairchild had found three thousand dollars for Popenoe to make one more expedition. Although Popenoe agreed with Galloway that America already had plenty of foreign plant material, he welcomed the assignment as a way to escape office politics and return to the tropical jungles he loved. On October 1, 1919, Popenoe left his comfortable routine in Washington—writing articles about his travels by day and socializing with his brother and pretty young women at night—and embarked on his fourth plant expedition to South America.

"I am not going to concern myself for the next year or so with much of anything but agricultural exploration," he told his parents on December 11, 1919, after he had settled in Guatemala City. "This will almost certainly be my last big trip and I would like to set a high-water mark for efficient exploration, so I can retire and live on my rep."

But this trip brought problems, too. In February 1920 he was detained briefly on suspicion of trying to overthrow Estrada Cabrera, Guatemala's tyrannical president. Government agents "caught me once in Antigua . . . and seemed to think they had a prize," he told Fairchild. "They went through me from head to foot and finally struck what they considered was going to prove damning evidence in the form of a Unionist (the revolutionary party) propaganda sheet. But it turned out to be a religious tract that a missionary had just given me, and they finally had to let me go, rather to their chagrin."[2]

A few weeks later, he had an accident while he was touring Zacapa, a rural area known for fruits and tobacco, and suffered a deep cut in his right foot. Doctors on the United Fruit Company's staff patched him up competently but without much sympathy. "They laid me on the operating table and washed their hands with alcohol and shot me full of cocaine," he told Fairchild in March 1920, "and then cut a big chunk out of the sole of my right foot and put four stitches in it, and then told me it didn't hurt."[3]

The accident occurred about the time Popenoe's mother died suddenly in California. Fairchild tried to cheer him with a good-natured note urging him to take care of himself and get back to work. "There is no use at all for an explorer on crutches. He's just a common ordinary employee, that's all," Fairchild wrote on April 18, 1920. "Watch that foot of yours as carefully as though it were the only avocado seed in the world and the whole avocado industry depended upon its growing."

Despite his injury Popenoe managed to ship 5,000 fresh avocado seeds to Washington. By May, he was well enough to move on to Costa Rica. Conditions there were rougher than they had been at the United Fruit Company facilities in Guatemala, and like Frank Meyer, Popenoe was forced to sleep in flea-infested beds. One night he recorded seventy-four flea bites on his left arm and eighty-six on his right. He longed to return to Guatemala. "Whenever I am there I think how good it will be to get out of it," he told Fairchild, "and when I am out of it I think how much I would like to be there."[4]

Despite the discomfort, Wilson Popenoe had many countries to explore before he could return to Guatemala. By summer 1920 he was in Colon, Panama. Next he took a ship to Colombia, where he chose Bogotá as his headquarters and made brief trips into the countryside.

Despite the tragic consequences of Frank Meyer working alone under miserable conditions, Fairchild couldn't afford to pay a team of explorers, as he had during Popenoe's first trip in 1913, so Popenoe grew lonely. By Christmas 1920 he had moved to Quito, Ecuador, where he had a rare opportunity to wear the tuxedo he always carried to a holiday dinner at the American Legation headquarters. The socializing was brief, however, and after more than a year on the road, the rigors of foreign plant exploring were wearing down even good-humored Wilson Popenoe.

"It's all right to spur up for a short time and when the circumstances demand it, but to work under pressure for several weeks is too wearing," he told his father the day after Christmas 1920. "There is no doubt that the principal cause of 'blueness' in myself, as well as with you, is nothing but fatigue. . . . I am such an old hand at this game now that I am not in any danger of getting really homesick. . . . I never get to the point where I feel really discouraged and want to quit. No sir, no son of my mother will ever quit!"

FIGURE 22. Wilson Popenoe on the equator in Malchingui, Ecuador, in 1921. Photo courtesy of Hunt Institute for Botanical Documentation, Carnegie Mellon University, Pittsburgh.

By then Paul Popenoe had married Betty Stankovitch and moved to California. Their engagement was so short that Wilson had never met his sister-in-law, so Fred Popenoe sent him photographs of Betty and her younger sister Emily. Wilson Popenoe, who always chased women when he could, developed a long-distance crush on Emily and pleaded with his father to tell him everything he knew about her. "The fact is, Dad, I am emotionally starved, horribly starved," he wrote on May 8, 1921. "You know how I have always been: always or nearly always in love with somebody. . . . I've got to love somebody, even though that somebody does not know it. I've got to have somebody to think about, to keep me from losing my morale down here in these frightful countries." He especially wanted to know how smart she was. "Has Emily a really good intelligence, one of the sort that will, in later years, read the *Atlantic* with appreciation?" he asked at the end of the letter. His infatuation led nowhere.

* * *

After Popenoe had explored five countries in Central and South American on foot and horseback and had collected thousands of avocado seeds, he began to worry about how well his material was being handled in America. The office was too disorganized and understaffed to handle properly all the material he sent in, he feared. By the summer of 1921 Popenoe told his father he didn't want to spend months looking for valuable plants just to have them die in America. "I am not going to break my back as I have been doing the past three months if there is not a good chance that everything which reaches Washington alive will get good attention," he complained on June 25, 1921.

By September, after an expedition to Peru and Chile, Popenoe had had enough. He told Fairchild that he was finished with plant exploring. "Never again, by all that is great and holy, do I tackle, alone, a two-years' exploration in which I attempt to cover pretty near all the inaccessible, unlivable and utterly impossible portions of the immense South American continent," he told his boss on September 15, 1921. "No, sir. . . . The peaceful comforts of a quiet life seem to me, just now, very attractive."[5]

By early November Wilson Popenoe was back at his desk at the U.S. Department of Agriculture, flea-free but harried by office politics and struggling to keep the office of foreign seed and plant exploration alive.

The fears he had shared with his father turned out to be well founded. The avocados he collected on his last trip didn't amount to much in America, and despite his hard work, none of the avocados he found were valuable. "We never secured anything as good as the Fuerte," Wilson Popenoe admitted in a speech in 1947.[6]

* * *

Fairchild, disappointed that Frank Meyer had died before exhausting China's plant possibilities, found a replacement as fast as he could. He was Jesse Baker Norton, a plant breeder and asparagus expert who already worked in the department. Norton was born in Tennessee, but like Fairchild and Swingle he had graduated from the Kansas State College of Agriculture. He was forty-two and a well-liked and respected colleague when he joined the ranks of Fairchild's agricultural explorers.

Norton set off from San Francisco on May 1, 1919, stopping in Honolulu, Vladivostock, Nagasaki, and Shanghai before arriving in Foochow, China, on June 22. He collected a few things—Chinese olives, a small watermelon, and a new lawn grass—but he wasn't happy. After only two months, he pleaded with Fairchild to bring him back to America. "He has gone all to pieces mentally and is so homesick that he is begging to come home at once," Fairchild told Orator Cook on August 20, 1919. Fairchild, probably remembering what happened to Meyer, quickly agreed to end the mission.

"Norton was convincing himself that he was not the kind of man to stand the strange, lonesome isolation of those regions where you are forever elbowing a crowd of inquisitive Chinese," Fairchild wrote later in an unpublished article. "As one of my friends familiar with China once remarked: 'A man must either be a missionary or a heavy drinker to stand the life in China.' Norton was neither."[7]

Norton's failure did not stop Fairchild. In March 1920 he found another candidate, this one a moody, self-taught botanist who had emigrated from Europe to the United States. He wasn't a missionary or a heavy drinker either, but he pleaded with Fairchild to send him to Asia.

He was Joseph Rock, a thirty-six-year-old Austrian with an unconventional background. Born in Vienna, Rock had had a difficult childhood. He was a sensitive boy who was raised by his domineering father after his mother died when he was six years old. A devout Catholic, Rock's father, Franz, had macabre demands about how to show respect for his dead wife. He ordered his heartbroken son to place a flower in her cold, lifeless hand during her funeral, a chilling experience young Rock never forgot. Franz Rock regularly took his son to her grave and ordered him to kneel beside it and cry.

Rock escaped his father whenever he could. As a student, he often played hooky and hung out with Arab beggars on the streets of Vienna, pretending to be an explorer on assignment in exotic lands. He claimed he had taught himself Chinese.

Rock's imagination had another rich source of inspiration. His father worked as a steward to a Polish aristocrat in Vienna, so the Rock family lived in the servants' quarters of the count's winter palace. Rock got a glimpse—from an uncomfortable distance—of the luxuries that came with wealth and power. By the time he was a teenager, Rock showed that he was brilliant, independent, and, if necessary, devious enough to get whatever he wanted.

After he finished school, Rock left Vienna and traveled. He spent three years wandering around Europe. (He later told American academics that he had been studying at the University of Vienna without earning a degree, an apparent lie.) Rock traveled across the Atlantic by working as a ship's steward. In New York he pawned his steward's uniform for 50 cents and took an aimless series of jobs, moving farther west every year. Eventually, at the age of twenty-three, he ended up in Hawaii, where he found work establishing a herbarium, even though he knew nothing about botany. "Arrogance, charm, and the fact that no one checked his credentials carried the day for him," explained S. B. Sutton, his biographer.[8]

It was a pivotal point in Rock's life. In Hawaii he discovered wild tropical plants that grew only in the islands. The territory's division of forestry hired him to collect specimens, a job that led to a position teaching botany at the College of Hawaii, then a small school with only twelve teachers.

During vacations Rock traveled alone through the Far East at his own expense, occasionally corresponding with David Fairchild about unusual plants he spotted. In May 1913 Fairchild made Rock a member of his global network of plant experts as a dollar-a-year collaborator.

While Rock was living in Hawaii he had learned about the plight of lepers sequestered on Molokai, a remote territorial island. The colony had a tragic history: leprosy was first noticed in the mid-nineteenth century after the disease had possibly been carried on ships crossing the Pacific. In 1865, the Hawaiian board of health ruled that having leprosy was a crime and, beginning in January 1866, ordered victims to be quarantined under miserable conditions. Eighteen years later, in 1884, the Hawaiian courts finally rejected the board of health's position and decreed that leprosy was a disease, not a crime. Although officials continued to exile the lepers, scientists and doctors finally began searching for ways to treat them.

Rock's boss, Arthur Dean, a chemist who was the College of Hawaii's president, was studying the therapeutic value to lepers of chaulmoogra oil, the product of a rare tree that grew only in the jungles of Southeast Asia. Dean told Rock about the oil, and during a trip on his own to Siam in 1919, Rock saw signs that it helped lepers living in an asylum in Bangkok.

Rock was a touchy, proud man. In the spring of 1920, he got into a dispute with his superiors about how to handle plant specimens he had collected. He could not get them to see the situation his way, so he quit his job at the college. Although brilliant, Rock's skills as a tropical botanist had limited value; his knowledge didn't apply to plants in temperate climates.

He left Hawaii for the United States on May 25, 1920, and for several weeks tried and failed to find another job. Finally, in July 1920 he arrived in Washington to meet Fairchild in person. Fairchild was impressed by Rock's scientific expertise and his independent travels in the Far East. Most important, he desperately wanted to replace Meyer in Asia. After a brief interview, Fairchild gave Rock an assignment: help Hawaii's lepers.

Wilson Popenoe, who was looking for avocados in Colombia when he learned about Rock's hiring, was pleased that Fairchild had managed to find a second plant hunter. "The office writes that they have put on a new explorer, Jos. Rock of Hawaii, who is a first-class botanist, has traveled a good deal and I believe will make good," Popenoe told his family on August 21, 1920. But he predicted the move would probably squeeze his own already tight budget. Popenoe was living frugally in a small Bogota hotel where he paid $3.50 a day for a bed and 25 cents for a cold bath. (Those prices would be about $40 and $3 today.) "I suppose with Rock going out the office will be poorer than ever and I will have a hard time to get any more money," he complained.

The Latin name for the chaulmoogra tree is *Taraktogenos kurzii King*, but Burmese natives called it the Kalaw tree. When he arrived in Southeast Asia in the fall of 1920, Rock had a clear mission from Fairchild: spend six months in Southeast Asia and collect enough seeds to start plantations in Hawaii and other U.S. tropical possessions. It was not an easy assignment because, despite what Rock had told Fairchild, he had no idea where to find the seeds. Nonetheless, Rock hunted for them with a style and sense of adventure that Fairchild admired.

Rock rarely missed an opportunity to boast about the risks he faced. Two months after leaving San Francisco, he was walking through the jungle near Kuala Lumpur when he saw a King Cobra, the world's largest venomous snake. It was "a huge beast, only about 15 feet away from me," he reported to Fairchild on September 26, 1920. "I walked somewhat carefully after that."

From Kuala Lumpur Rock took trains to Bangkok where, as required by plant exploring etiquette, he introduced himself to the new United States minister, George W. P. Hunt.

Hunt, a Democrat who had served as Arizona's first governor, had gotten the job in a nontraditional way. A jovial three-hundred-pound man with elaborate whiskers, he called himself "Old Walrus," although an associate said he looked more like Buddha with a handlebar mustache. Earlier in 1920, when Hunt was considering challenging Arizona's incumbent Democratic U.S. senator, the senator's supporters asked President Woodrow Wilson to stop the primary by appointing Hunt to a distant foreign post. Wilson took a globe, pointed to Siam, and asked, "Is this far enough?"[9]

Quickly whisked out of Arizona, Hunt found himself with a fancy title—Minister Extraordinary and Plenipotentiary—but without any diplomatic skills or interest in his new job. Rock, who had learned as a boy to appreciate refinement and social graces, was appalled by Hunt's informality. The American diplomat did not behave like the European royals Rock had watched in Vienna.

"When sending in my card I was ushered into a large room to a central desk at which sat a corpulent man, with a round red face and a yellow

FIGURE 23. George W. P. Hunt, U.S. ambassador to Siam, in 1920. Dr. Joseph F. Rock/National Geographic Creative, image ID 605085.

mustache, the ends of which were waxed to two sharp points," Rock wrote in an unpublished manuscript. "He greeted me in a very democratic manner, took the letter of introduction and at the same time motioned me to take a seat opposite him at the long desk. He smoked cigarettes and opening a silver cigarette case, said, 'Have a cigarette.'"[10]

After staying in Bangkok for a month with Ambassador Hunt, Rock left for Chieng Mai, where his search for chaulmoogra seeds began. Rock wasted little time informing Fairchild about his situation. "I am writing you from paradise, for indeed this place is worthy of that name," he reported on October 29, 1920. "I am camping in the mountains north of Chieng Mai in a wonderful forest, beside a little stream which sends me to sleep like a lullaby."

Three days earlier Rock had climbed a 5,500-foot mountain and spied from the top a spectacular vista that stretched into China, his first glimpse of the land he had dreamed about as a boy. "Mountain after mountain and range after range, a perfect virgin field untrodden by any botanist or agricultural explorer," he described the scene on October 29, 1920. "It is a real paradise for a plant lover to be in a pine forest and to look down thousands of feet into deep ravines and forest-covered ranges inhabited by tigers, panthers, and elephants."

Unlike Meyer, Rock did not travel alone or simply. On his hunt for the chaulmoogra seeds, his retinue included nineteen laborers, a cook, an interpreter, and a boy (job unknown). He didn't have a sedan chair, but he did ride a pony given to him by a village dignitary, and local workers carried his belongings on three ox carts.

In late December 1920, after he crossed the border into Burma, Rock finally spotted a chaulmoogra tree near Moulmein, but he couldn't collect seeds because the tree wasn't bearing fruit. Natives told him more trees grew farther along in a village called Oktada. He traveled by train and ox cart and, toward the end of the journey, on foot. He moved in a crouch through thick bamboo forests across ground that showed tiger tracks. Rock found the trees, but they weren't right either. Although they looked almost identical, they were not true Chaulmoogra trees and would not produce seeds with the powerful oil.

Undaunted, Rock returned to Rangoon for better information. Officials there told him the true chaulmoogra tree grew only in the upper Chindwin River valley. Rock set off again and traveled first by train to Monywa. There he found a batch of authentic Chaulmoogra seeds on sale in the village market, a loathsome place. "The bazaar is a living entomological collection," he reported in the *National Geographic Magazine*. Flies "cover the conical piles of brown sugar spread out on mats on the ground to such an extent that almost every grain is moving, and this in the midst of squatting, betel nut chewing, and expectorating women."

The seeds he wanted, someone told Rock, came from "up north." He kept moving: one more leg by boat, another by dugout canoe. He enlisted twenty villagers to carry his gear and led them on a two-day march to a village called Khoung Kyew. There he found one genuine tree, but it had no fruit. The village headman suggested trying another place deeper in the forest. This was Kyokta, a settlement of only thirty houses surrounded on three sides by thick jungle. Rock and his group marched another five miles along a dried creek bed and finally, seven weeks after Rock left Ambassador Hunt in Bangkok, found themselves in a forest of chaulmoogra trees, covered in fruit. Their timing was perfect.

FIGURE 24. Villagers in Kyokta, Burma, killed this man-eating tiger by spearing it twenty times. Photo by Joseph F. Rock/National Geographic Creative, image ID 605097.

Rock and his workers—his entourage had grown to thirty-six—spent hours collecting seeds. On their return to Kyokta they spotted more tiger tracks in the creek bed and realized that an animal was on the prowl. "It made me rather uneasy, as we had to pass through dense jungle thicket, and in my imagination I saw [the tiger] lurking in every dark spot," Rock admitted to Fairchild on January 23, 1921.

Rather than risk an encounter with the animal on the way back to Khoung Kyew, the group stayed in Kyokta. They were too exhausted from their labors to hear the commotion in the middle of the night when the tiger attacked. It entered a hut in a nearby rice field where three women, a two-year-old girl, and a five-year-old boy were sleeping. The tiger instantly killed the little girl and two of the women. It badly injured the third woman and mauled the boy and tossed him, still alive, into a campfire. Then the tiger stalked into the forest. The boy managed to run into the village to report the attack.

The villagers—including one man who was the father, brother, and husband of the victims—immediately built a trap for the animal. They baited it with one woman's body and waited. A huge thunderstorm hit in the middle of the night, rousing wild elephants to stampede through the village. Despite the commotion, the trap worked.

The ferocious tiger was tied up when Rock arrived in the morning. Furious villagers killed it with more than twenty spears while Rock took photographs. After this incident Rock burned bamboo every night as protection because the shoots exploded like gunshots and scared wild animals.

Rock immediately relayed the news about this adventure to Fairchild, who was in Washington trying to keep his superiors interested in his romantic project. Fairchild quickly contacted Gilbert Grosvenor to learn if Rock could tell his story in the *National Geographic Magazine*. "Any publicity we get now may help us in our attempt to keep our work from going to pieces because of a lack of appropriations, which looks pretty serious to us," Fairchild wrote on January 28, 1921.

Grosvenor was willing to help his brother-in-law. "I am always too glad to cooperate with you in making your important work known to the American public," he answered ten days later. By then, Grosvenor had made the magazine an enormous success; about 750,000 society members received each issue.

Rock's account appeared in the March 1922 edition. It was illustrated by many photographs, including one of the dead tiger, and immediately caused a sensation. One society member was Joseph Nelson Rose, a botanist who had explored Mexico for the Smithsonian Institution and had briefly rented a room in his Washington home to Wilson Popenoe. The article astonished Rose. "Joseph Rock was the greatest explorer we had ever known," Rose told Popenoe,[11] who was offended because he wanted that title for himself.

Rock's adventures didn't impress everyone. After he escaped from the jungle and returned to America, bookkeepers at the office of seed and plant introduction determined that he had spent too much money hunting for chaulmoogra seeds. Although Fairchild had given Rock a bigger budget than Popenoe, the accountants asked Popenoe to forfeit $600 of his own expense account to pay Rock's bills. Popenoe grumbled to his boss about the injustice, but Fairchild ignored his complaints.

Rock was proud of his success. Grosvenor paid him $400 for his article and photographs, a large fee (about $5,400 in current dollars) that allowed

Rock to celebrate by visiting his sister and her family in Austria. In August 1921 he returned triumphantly to Vienna as a real explorer from faraway lands.

Fairchild was also proud of Rock's work, which he described as "experiences quite as thrilling and dangerous as any to which explorers in tropical countries are liable, including a unique one with a man-eating tiger."[12]

* * *

Rock's first expedition had gone so well that Fairchild sent him back to Asia four months later. This time Rock was assigned to enter China, the land he had dreamed of exploring since he was an unhappy child in Vienna. In December 1921 Rock left New York City on the trip that would change his life.

His destination was Yunnan in the southwestern edge of the country on the Burmese border, the province reputed to have the richest plant life in the world. On February 1, 1922, Rock was on the Burmese frontier. "I shall soon cross into mysterious China," he reported to Fairchild.

Despite his anticipation, Rock's first glimpse was unexciting. "At about 5 p.m. we reached the Chinese-Burma border at Chieng-Law," he recorded in his diary. "A much faded Chinese flag was implanted almost in the center of the road. To the left of it was a bamboo-wooden shanty where the Chinese official, a small unwashed fellow but with a kindly smile, gave us a rather nice reception."[13]

The trip's primary purpose was to finish an assignment that Frank Meyer had started before he died: save the American chestnut tree.

At the beginning of the twentieth century, America's native chestnuts thrived along the Eastern Seaboard. An estimated 4 billion trees—many as tall as one hundred feet—covered about a quarter of the region's forests. Chestnut wood was hard and straight and vital to serve the nation's growing needs for railroad ties and telephone poles. But in 1904 a scientist at the Bronx Zoo in New York City noticed a canker or fungus spreading on the trees' bark. Three years later the same disease was evident on chestnut trees growing across the street in the New York Botanical Garden. It was the beginning of the most significant invasion of a foreign plant disease in American history.

In January 1907 a member of Congress briefly questioned a scientist in

Charles Marlatt's office about the problem. As always, the divisions and rivalries at the agency were clear: Marlatt's Bureau of Entomology was concerned only with insects; Beverly Galloway's Bureau of Plant Industry was concerned with plant diseases. The scientist dismissed the question in one sentence. "In studying the insects we always determine whether or not there is some other primary cause," explained Andrew Delmar Hopkins, a forest insect investigator, "and if we think it is caused by a disease we turn the subject over to the Bureau of Plant Industry, confining our work entirely to the investigation of insects."

Little progress was made to combat the epidemic for several years, although it continued to spread, in the words of Susan Freinkel, an expert on the disease, wrote in *American Chestnut*, "as a smoldering wildfire that lurched forward by sending out spores in advance."

Finally in February 1913, after scientists in the agriculture department concluded that the fungus might have been introduced from Asia, Fairchild had asked Meyer to look for signs of the disease. Only four months later, in the mountains north of Peking, Meyer found the fungus on Chinese chestnut trees, but it was a mild case, indicating that those trees had a natural ability to withstand the disease. Scientists hoped they could use the foreign varieties to breed immunity into domestic trees.

Meyer mailed a two-inch square of infected bark to Washington, which allowed scientists to concoct inoculations, but the protection didn't stop the bark fungus, which was inexorably wiping out America's forests. Scientists have never determined exactly how the fungus was introduced, but many believe it was carried by chestnut trees imported from China and Japan at the end of the nineteenth century.

The death of America's chestnut trees was so serious that Fairchild later admitted that it softened his resentment of Marlatt and his staff. "I felt like saluting the pathologists who are working to prevent a repetition of this calamity in the future," he wrote in his autobiography. "I regretted any feelings of impatience I may have had towards their quarantines and inspections."[14]

* * *

For the rest of his travels, Meyer diligently collected chestnuts in China and Japan and sent dozens to Washington, but none made much difference.

After Meyer died, Fairchild still hoped to find the chestnut's savior, and he asked Rock to penetrate deeper into China to find more specimens.

By May 1922, Rock was settled in Likiang, a city in Yunnan province. He told Fairchild that it was a botanist's paradise and, unfortunately for Rock, it was filled with botanists.

George Forrest and Frank Kingdon-Ward, celebrated British plant explorers, worked nearby. Rock, feeling flush with the U.S. government's money, hired two of Kingdon-Ward's workers while their boss was away botanizing. Kingdon-Ward, who was angry about Rock's impudence, wrote soon afterwards that plant exploring had become too popular. "When I joined the ranks, plant hunting was still an adventure; now it is a trade," he said. "Presently it will be a trade union."[15]

Rock was careful to treat Fairchild with more respect. He wrote long letters to his boss filled with descriptions of his beautiful surroundings. He also shipped many parcels of seeds and plants to Washington. But few turned out to be useful.

Despite the office's budget constraints, Fairchild had found $6,000 for Rock's second expedition, the same amount budgeted for Meyer's last trip. The money was supposed to last for two years, but after only a few months Rock asked for more, a demand other staff members resented.

Desperate to keep alive his romantic mission of scouring the globe for new food and plants, Fairchild defended Rock and his extravagant ways, even when the criticism came from his protégé, Wilson Popenoe. But the pressures on Fairchild were beginning to show, making him temperamental.

"One day he sort of went up in the air and told me . . . that I was jealous of Rock," Popenoe complained to his family on January 30, 1922. Rock "spends like a drunken sailor, giving dollar tips and all that sort of thing. Everyone in the office except DF is dead set against him. DF says if he gets the goods we can afford to put up with his idiosyncrasies."

It soon became clear that Rock's ways were not those of the federal government's. Once he bought thirteen horses in Yunnan and dutifully included the purchase on his expense account, but when he sold them he didn't bother collecting receipts.[16]

Rock lived well, even in Yunnan province. He carried a portable record player so he could listen to Enrico Caruso in the evening. He taught his

FIGURE 25. Joseph Rock often wore traditional Tibetan clothes while he worked in Yunnan. Photo from the Arnold Arboretum Horticultural Library of Harvard University, © President and Fellows of Harvard College. Arnold Arboretum Archives.

Chinese cook to make rich Viennese dishes. To remind himself of his sad childhood, he carried a copy of *David Copperfield* by Charles Dickens.[17] Later, when he had a bigger budget from a new employer, Harvard University, Rock bathed regularly in a portable bathtub he bought at Abercrombie and Fitch in New York City.

Expenses like that added up. By the end of 1922, Rock had spent all his money, so he again asked Fairchild to send more so he could remain in China for an extra year. The request was more than the U.S. government— or David Fairchild—could take.

*   *   *

Fairchild desperately wanted Rock to continue exploring, yet he knew he couldn't get enough government money to keep him in China. For the first five months of 1922, Rock had spent about $7,000 (about $95,000 in current dollars), way over his budget. Without first discussing the plan with Rock, Fairchild persuaded Gilbert Grosvenor to finance Rock's next expedition. The National Geographic Society had become enormously influential—the organization had already sponsored important expeditions, including Robert Peary's trip to the North Pole in 1909—and Grosvenor, as supportive as ever, agreed to help. Through the society he raised $17,000, a large increase over Rock's government salary and expenses. Nonetheless, the high-strung explorer was upset when he found out he had been transferred from the U.S. government to the National Geographic Society. "I must say I do not like being unceremoniously shifted that way," he complained to Fairchild on January 30, 1923. "Why could I not have been given leave without pay for the period I am to work for the National Geographic rather than be dismissed? I do wish to know where I stand."

Characteristically, Rock threatened to resign. In the same letter, however, his tone suddenly switched from defiant to pathetic. "Now please, my dear, dear Dr. Fairchild, do not misunderstand me," Rock continued. "You do not know how much I really love you like a son his father and I do want to have the great pleasure of working with you for a long, long time to come. . . . I cannot believe that you wish me to leave your office after this year."

Fairchild wrote friendly, patient letters to calm Rock down and assure him he could continue to send plants to the agriculture department, even though his government salary had been cut back to a dollar a year. Unlike Fairchild, however, Popenoe and others in the office were glad to see him go.

"Rock has to be handled like a grand opera or movie star," Beverly Galloway advised on March 26, 1923. Galloway complained that Rock had not sent much valuable material. Even his images were virtually useless to the practical needs of the department. "A day or two ago I was looking at his photographs, a stack nearly six inches high," Galloway wrote. "I should say that nearly ninety percent of them are wonderful mountain views, including snowcapped peaks, alpine meadows, beautiful vistas, some forest scenes, wonderful awe-inspiring gorges, groups of wild people, etc. There are not more than a dozen plant photographs in the lot."

Rock had found more chaulmoogra seeds, but they apparently had no value. Although a plantation had been established on the island of Oahu from Rock's original shipment from Burma, it was becoming evident to doctors that chaulmoogra oil did not cure leprosy. The medicine was difficult to administer, resulted in painful side effects, and was often ineffective. Eventually doctors turned to antibiotics to treat lepers.

Washington officials were especially disappointed that Rock hadn't found many chestnuts. "He landed quantities of seeds, though with only partial success," Fairchild wrote in an unpublished account of Rock's work, "for chestnuts are among the hardest seeds to ship, and he was forced to send them by coolie across the mountain passes, from which cold, high altitudes they had to cross the Indian Ocean, or else make the long journey down the Yangtze River."[18]

Only six chestnuts from Yunnan arrived alive in Washington. By February 1, 1923, Rock officially switched to the National Geographic Society. By 1950, virtually every American chestnut tree had died from the mysterious Asian disease.

proof

Grumpy Old Bachelor Tramp

Frank Meyer, worried about what he would do if he retired from the department, had managed to save a good deal of his salary and had invested the money well. At his death he had an estate valued at $28,446.86 (almost $500,000 in current dollars), enough money for him to make a grand gesture.

After providing for his family, Meyer bequeathed $1,000 to his friends and colleagues in the office of seed and plant introduction so they could have another bonfire party in his honor. "It isn't much, but I feel grateful to them for all they have done for me," he told Fairchild after he wrote his will.[1]

After Meyer's death, however, Fairchild decided to put the money to permanent use. He established, under the auspices of the American Genetic Association, the Frank N. Meyer Medal to be awarded for distinguished service to foreign plant introduction. And there was no doubt about who would be its first recipient. On May 3, 1920, at a short ceremony in an office in downtown Washington, Fairchild presented the handsome bronze medallion to Barbour Lathrop.

Fairchild had designed the medal with great care. One side showed an ancient Egyptian image of men loading incense trees onto a ship. This was a copy of a bas-relief in Queen Hatshepsut's palace in Thebes that Fairchild and Lathrop had visited in 1901. The other side bore illustrations of a Chinese jujube and a white bark pine tree—both Meyer introductions—with a Chinese inscription that means "In the glorious luxuriance of the hundred plants he takes delight."

FIGURE 26. Barbour Lathrop is awarded the first Frank N. Meyer Medal for distinguished service to foreign plant introduction on May 3, 1920, in Washington. From left to right: Walter Swingle, Beverly Galloway, Mark Carlton, David Fairchild, Peter Bisset, Barbour Lathrop, Thomas Kearney, William Taylor, Orator Cook, Edward Goucher, and Howard Dorsett. Photo by M. M. Blaine. Photo courtesy of Hunt Institute for Botanical Documentation, Carnegie Mellon University, Pittsburgh.

Lathrop was seventy-three when he accepted the Meyer medal. By then, his traveling had slowed down, but it had never stopped. He had made many trips since August 1903, when he and Fairchild had arrived in Boston after their last voyage together.

While Fairchild hurried south to Washington to look for work, Lathrop traveled north to York Harbor, Maine, for his regular summer vacation with his sister, Florence Page, and her family. It was an annual stop in Lathrop's predictable routine.

The next year, as Fairchild settled into his life in Washington, Lathrop resumed his almost-incessant foreign wandering. On one trip he brought with him Drummond MacGavin, the son of a friend who worked as the cashier in Lathrop's San Francisco bank. MacGavin, twenty-two, was an agreeable companion, and Lathrop looked upon him almost as a son. Nonetheless, he admitted it wasn't the same as traveling with David Fairchild. "I always recognized its interest, but never dreamed I'd so sadly miss the daily plant talk, the collecting, the shipping and the news of success or failure," Lathrop wrote to Fairchild from his oceanfront room in the Galle Face Hotel in Colombo, Ceylon, in 1904. The "sight of quaint places and peoples [formerly] made travel an unwearying pleasure. Now it fails to satisfy me as it did. . . . I don't acknowledge it or permit myself to show it, but much of the keen interest is gone."[2]

Fairchild, who maintained a lively correspondence with his mentor, grew more tolerant of his crusty personality as the two men grew older. When he fell in love with Marian Bell, he immediately telegraphed Lathrop about the engagement. Lathrop, who was staying in Shepheard's Hotel in Cairo when he got the news, was exuberant. He danced around the room in an undignified manner and let loose with "several cowboy whoops," he told Fairchild on March 18, 1905.

The engagement gave Lathrop a new subject for advice and counsel, all unbidden but not unwelcome. Although he was single, Lathrop had strict rules about how to be happily married. "Don't worry your wife with every trifle of your daily annoyances, but be sure you make her full partner and adviser in all the serious affairs of life," he announced in the same letter. "Just be good comrades and your lives will really become almost one." He said he had only one regret about Fairchild's marriage: "No more tramps together." Lathrop, who left few clues about his personal life, avoided weddings and managed to be in Naples when the couple married. He sent them a check for three hundred dollars.

Shortly after their wedding, Fairchild took Marian to Boston to introduce her to Lathrop. Their first meeting was difficult. When they entered Lathrop's suite in the Hotel Touraine, Fairchild said, "Uncle Barbour, this is Marian."

Lathrop looked her over and remarked, "Well, she has fine eyes, at any rate."

Next, cutting off Marian before she could say anything, he turned to Fairchild and complained that his hair was too long. He accused Marian of trying to make him look like her father, the eccentric inventor with long, wild hair. Marian, devastated, burst into tears and left the room.[3]

Despite the insult, she returned quickly and with grace admitted that perhaps it was time for Fairchild to see a barber. The new friends got along well after that meeting. David and Marian named their first child in honor of Alec Bell, but their second one, Barbara Lathrop Fairchild, honored him.

Over the next decade, Lathrop kept sailing around the globe, making regular visits to Maine and San Francisco when he was in America. Various letters and diary entries offer glimpses into his fashionable life. According to Thomas Nelson Page, his brother-in-law, on a fine spring day in Paris in 1906 Lathrop ate lunch at a chic, notorious café called Le Foyot near the Luxembourg Gardens that had been damaged a few years earlier by anarchists who placed a bomb in a flowerpot. A week later, Page and Lathrop dined together at the Hotel Ritz in Paris.

Lathrop was always generous. In about 1913, as Lathrop was leaving alone for another voyage to the South Seas, Fairchild asked him to continue to send new plants to Washington. To facilitate these shipments, Fairchild appointed him as a collaborator with the U.S. Department of Agriculture at a salary of a dollar a year. Despite the miserly pay, it was an official position, and Lathrop had to fill out a government questionnaire to get the job. He said he was careful to answer each question properly, until he got to the end. "Last query of all was somewhat to this effect: 'What would be your occupation or work during the hours in which you were engaged in this position?'" he said later. "And my answer was, 'Spending my own money for the benefit of the department.'"[4] He never cashed the annual one dollar checks, preferring to save them as souvenirs.

He called himself "a grumpy old bachelor tramp."[5] As he aged, he grew wealthier but sicker and unable to take many foreign trips. Every year, however, he managed to travel across America by train and to shuttle up and down the Eastern Seaboard with a car and driver.

As an active Bohemian Club member, he loved to participate in the annual summer High Jinks retreat at the club's camp in the woods outside San Francisco. One of the club's best-known members was George Sterling, a respected but troubled poet. Because Sterling was broke, Lathrop secretly paid the rent on his rooms at the club for more than a decade.[6] Although he was apparently unaware of Lathrop's generosity, Sterling dedicated a long poem called *Lilith* to Lathrop. The patronage ended in November 1926 when Sterling killed himself in the room by swallowing a cyanide pill.

After the summer festivities in California, Lathrop always traveled to Maine to visit his relatives. His great-nephew Henry Field remembered that Uncle Barbour's regular visits followed a familiar pattern. He always arrived in a shiny, chauffeured Cadillac, fastidiously dressed in a white suit, with black shoes and socks. He wore a light-colored tie held in place by a black pearl stickpin, as well as a wide-brimmed Panama hat, and carried an elegant cane made of tropical wood. He traveled with several trunks and a tin sea chest labeled, "Barbour Lathrop, The Bohemian Club, San Francisco, California."

He brought gifts from afar for everyone and regaled his dinner companions with stories about his adventures, although they had heard them many times. He usually stayed with his family for five days every August.[7]

* * *

A new stop was added to his itinerary in 1916 after David and Marian Fairchild bought an old estate on Biscayne Bay in Coconut Grove, Florida. A house and a few outbuildings filled the eleven-acre property that, in deference to their adventures in Java, Lathrop christened as the Kampong, the Malay word for village. Fairchild landscaped the property with seeds and plants from around the world and created a lush, semi-tropical enclave.

Lathrop visited each winter to hang out with Fairchild's family and friends. The Coconut Grove crowd included Marian and David and their three children—Lathrop called them his "Fairchildren"—and Elsie and Gilbert Grosvenor, who bought the property next to the Kampong and often entertained adventurers connected with the geographic society.

Alice Barton Harris, Marian's school friend from Washington, and her husband, Franklin Harris, who was called Harry, lived nearby, too. The Harris's next-door neighbor was Marjorie Douglas, a young writer whose father had started the *Miami Herald*. It was a lively, unconventional group with lots of interesting things to talk about.

In 1917 Fairchild found another way to honor Lathrop's work for the U.S. government, although as usual Lathrop had to pay for it himself. As mangosteens and mangos had always been Fairchild's favorite introductions, Lathrop's choice was bamboo. He had marveled about its many uses in Asia, from construction timber to soup ingredient, and persuaded

FIGURE 27. David Fairchild and Barbour Lathrop seated on the banks of the Salinas River, near Bradley, California, on October 4, 1919. Photo by Marian Fairchild from the David Fairchild Collection, Fairchild Tropical Botanic Garden, 2876.

Fairchild to try to make it popular in America. "It is strange to consider that on one side of the Pacific there are civilizations comprising hundreds of millions of people who are so dependent upon the bamboo that they simply cannot imagine an existence without it," Fairchild recounted, "whereas, on the other shore, a hundred million people live whose main contact with the plant is through its use as a fishing pole."[8]

Early in 1917, a privately owned forty-six-acre bamboo garden on the Ogeechee River near Savannah, Georgia, was in danger of being abandoned. Fairchild had no money in his budget to save it, so he persuaded Lathrop to buy the property—it cost $5,000—and rent it to the government for $1 a year. Lathrop refused to allow the garden to bear his name while he was alive, but he supported it financially and stopped there every year on his drive from Florida to Maine. Fairchild dutifully kept him informed him about the garden, especially after the bamboos finally began to flourish. "It's such fun being in on the beginning of things, isn't it?" he asked Lathrop on July 20, 1921.

By then Lathrop's days of foreign travel had ended. In 1924 when Fairchild was going on a long trip without him, Lathrop didn't try to hide his sadness. "Give me a bit of a sketch of what is planned, for I love to mentally follow you, as I cannot bodily accompany you," he asked on September 2, 1924. "I want to feel that I am accompanying you in the spirit during every mile of your journeying."

Marjory Douglas, who spent many hours with Lathrop in Florida, saw how painful it was for him to remain in America while his protégé continued to visit remote corners of the globe. "It was a blow to Barbour Lathrop," Douglas recalled. "His letters were full of hearty congratulations and his pleasure in the opportunity for Fairchild and his constant interest in the object of their voyages. But it was clear he was torn by unhappiness that he could no longer lead such an expedition himself."[9]

Before Fairchild left America, Lathrop mailed him a check that he spent on a small camera, a thoughtful gesture that Fairchild said meant Lathrop actually would accompany him. "So, dear Uncle Barbour, you will be in my pocket, so to speak, and whenever I snap a scene it will remind me of you and you shall see vicariously every strange sight which your camera takes," Fairchild told him from Paris on November 28, 1924. "Thank you so much for the gift, you precious old friend."

When the Fairchilds were away, Lathrop stayed in Florida with their friends. Once a week, he invited Douglas, who was in her mid-thirties, to dinner at his hotel. He was old and frail but still crotchety. Meals had a predictable routine. "He chose as his own a table in the corner of the large dining room with his back to the wall and the window at his right hand firmly nailed shut, so that no officious idiot could open it and let a draft on him," Douglas remembered in *Adventures in a Green Land*. "He had the orchestra as firmly moved to the back of the dining room so that it would not interrupt his conversation. If the waiter irritated him, he had him removed, at the top of his voice. He had even been known to take his stick to a man he considered impertinent whom afterwards he always tipped heavily."

Lathrop enjoyed her company very much. "Marjory Douglas is a bounding ball of buoyant babyhood," he reported to Fairchild on February

FIGURE 28. Marjory Stoneman Douglas poses with her cat in the doorway to her home in Coconut Grove, Florida. Photo from the Marjory Stoneman Douglas Papers, courtesy of Special Collections, University of Miami Libraries, Coral Gables.

17, 1925. "I am very fond of that young woman and love the way in which she smilingly faces the world."

Eventually, however, even Lathrop's storytelling had to stop. Parts of his body became paralyzed, and he had trouble speaking. "It was dreadful," Douglas remembered. "You couldn't understand a word he said.[10]

In January 1927 his Florida friends tried to cheer him up by celebrating his eightieth birthday. "The party went off very well," Alice Harris, Marian's friend and neighbor, wrote in her diary. "He's very old and frail and pathetic. I wonder how many more birthdays he'll have!"

The Fairchilds, who were in Africa, didn't attend the party, but they often sent Lathrop colorful accounts of their travels. He loved reading them. "No letters I get from anybody please me more than those from you," Lathrop wrote on February 18, 1927. "Their coming makes a bright spot in a somewhat dull existence and I'd love to have the spots show up more frequently."

He grew feeble that winter and finally confessed in his February 18 letter that he was unable to do much. "As you would like me to tell you something personal," he admitted, "I am in a rather weakened condition, which prevents anything in the way of exercise beyond automobiling, and not too much of that, but I still can and do keep smiling."

Fairchild answered Lathrop's last letter with a cheerful tribute to the adventures they had together. "The years have flown since we were together, dear Uncle Barbour, and exploring the Pacific Islands in the little boat which Hughes got for us, but though all sorts of experiences have come into my life since those days none have come which were fuller of a strange romance than they," Fairchild wrote on March 23, 1927. "What a life you have had, dear Uncle Barbour. What thrills! What excitements! What glorious sensations!"

At the end of April, Lathrop was staying in the Bellevue Stratford Hotel in Philadelphia while he received treatments for one of his many ailments. Although he was fading quickly, Henry Field, his grandnephew, and Florence Lindsay, his niece, had time to get there. Years earlier Lathrop had told Fairchild that he was not afraid to die. "I have had a jolly good time in this world—somewhat shaded here by sorrow, there by regret, and there by physical pain," he wrote, "and hereafter, in the great somewhere, my soul will surely float from place to place, as my body has done here. The world wanderer will become the celestial tramp."[11]

Lathrop died in his room at the Bellevue Stratford on May 27, 1927. No one knew exactly how many times he circled the globe during his lifetime. When he was asked, he usually replied, "Eighty-three." But his friends guessed it was closer to twenty-six.[12]

In his hotel room after he died, his family found an assortment of personal belongings suitable for a world-traveling gentlemen, including an eighteen-carat gold watch, gold and pearl shirt studs, and stick pins of pearl, diamonds, and jade. They also found one unusual item: the bronze Meyer Medal that honored his work as, in Fairchild's words, "the patron saint of plant exploration."

Pushing On

After David Fairchild turned Joseph Rock over to his brother-in-law, he no longer had any plant explorers on his staff. Outsiders continued to send interesting seeds and cuttings, but the work wasn't the same. Federal officials' interest in the project cooled during the First World War, and Fairchild lost his most influential champion when Alexander Graham Bell died on August 2, 1922.

The following January, Fairchild, exhausted from fighting the government bureaucracy, went to Coconut Grove for the winter. By then Mabel Bell had also died, five months after her husband, leaving Marian Fairchild her share of the family fortune. Although it is not known how much money was left, the Bells had substantial assets that Marian shared with her sister, Elsie Grosvenor. "I have no doubt DF is well fixed for the rest of his life," Wilson Popenoe told his family on January 28, 1923. "And it has come at just about the right time for him, since his health is getting such that he can't do much more real work here in the office."

Fairchild stayed in Florida longer than usual in the winter of 1923 so he could establish a new government experiment garden at Chapman Field, a former airfield south of Miami. It was the agriculture department's third garden there. Hurricanes and real estate development had destroyed Swingle's first one on Brickell Avenue and a second one nearby called Buena Vista. The first trees were planted at Chapman Field (now officially called the Subtropical Horticulture Research Station) on April 26, 1923. Fairchild left Wilson Popenoe in charge in Washington while he was in Florida.

FIGURE 29. David Fairchild as a mature man. By permission of the Fairchild Tropical Botanic Garden, 6804, David Fairchild Collection.

At first Popenoe was pleased and proud to be running the office, but the longer Fairchild stayed away, the keener his absence was felt. "The Chief," as Popenoe called him, had created and sustained American plant exploring for more than twenty-five years. Fairchild's charm, his contacts with foreigners, his diligent labors, and most important his conviction about the value of sending brave explorers to exotic corners of the world had driven the enterprise. David Fairchild personified American plant exploring.

His friend Thomas Barbour said the scheme had succeeded for so long because of Fairchild's charm. "No more lovable human being every lived," Barbour wrote in his autobiography, *Naturalist at Large*. "The fact that he later traveled all over the world for years on end simply meant that he left a stream of friends behind him on every continent."

Popenoe, as Fairchild's loyal deputy, had a hard time keeping the operation going. After a year as acting chief, Popenoe admitted he was discouraged. "I seem to feel that the office is getting clericalized and it will be my own aim to counterbalance this tendency in so far as possible," he confessed to Fairchild on April 5, 1924. "I am going to keep from getting clericalized myself and hang onto the romance of the work. . . . I want to keep alive, so far as I can, the personality of this office, something of which I have always been very proud, something which you have instilled in the office for twenty-five years and which shall not die if I can help it."

While Popenoe was struggling in Washington, Fairchild was enjoying comfortable weather in Florida and organizing another romantic adventure for himself. Years earlier, Fairchild had met Allison Armour through Alec Bell. Armour, like Barbour Lathrop, was a wealthy philanthropist from Chicago interested in science. Around the turn of the century, Armour sponsored a few scientific expeditions and collecting trips in North Africa and Mexico's Yucatan peninsula.

Armour spent winters in Florida, where he kept a comfortable houseboat on a dock near Coconut Grove. In January 1922 he had invited David and Marian Fairchild to dine with him on board. Lathrop was another guest. After the meal Armour announced that he was bored in Florida and wanted to take a long, leisurely trip to explore for plants. Apparently nothing came of his comment right away because he repeated it a year later, this time as a more definite plan, and asked Fairchild to organize the voyage. Armour's invitation verified Wilson Popenoe's observation years before that his chief "certainly has a way of falling in with the big guns."[1]

The proposal excited Fairchild. "For the second time in my life I had met an expert traveler and a great gourmet who was interested in furthering the cause of plant introduction with which I had been so long associated," he recalled.[2]

* * *

Many things had changed since Fairchild began his odyssey with Lathrop. Now he was fifty-five and exhausted. There were signs all around that his cosmopolitan attitude toward foreign foods and useful plants had little support in Washington. The clearest signal came from the growing controversy over immigration, a national debate that culminated in Congress enacting the first quota system in May 1924. The law reflected a growing national attitude of isolationism and xenophobia.

And five months after Congress enacted the quota system, the office of seed and plant introduction stopping publishing *Plant Immigrants,* the bulletin that informed farmers about new plants available for experiment. The demise of the bulletin, supposedly to save money, crushed Fairchild. He had written most of its articles himself and had lovingly overseen almost every one of its 219 issues, many of which had full-page photographs submitted by botanists around the world. Its end was another sign that David Fairchild's romantic plans about international cooperation were doomed. "I tell you, dear Bert," he wrote to his brother-in-law, "it has been a struggle all of my life to keep what I have built up for Uncle Sam from being destroyed by the vandal functionaries."[3]

Fairchild did not hesitate to accept Armour's offer. "I felt I had done some of my best work as a plant explorer and, on the other hand, had balanced my years of travel by an equal number in an office chair," he wrote in his autobiography. "Our children were reaching ages of some discretion, and Marian and I began to yearn for the open road once more." Nonetheless, he did feel some guilt when he told his colleagues in Washington in 1924 that he was leaving. "I could see distinct disapproval in many faces," he recalled later. "They felt that I was once more 'abandoning the ship' as I had seemed to do in 1898 when I went off with Mr. Lathrop."[4]

When Fairchild departed on the long voyage with Allison Armour, he hadn't intended to give up his government job permanently. He kept his title, his full salary, and a five hundred dollar expense account, yet circumstances changed while he was away. Even his loyal deputy gave up trying to fight the new attitude in Washington.

Six months after *Plant Immigrants* ceased, Popenoe decided that his new bosses cared more about efficiency and organization than foreign foods. "Things in the bureau have reached the point where it seems useless for me to hang on any longer," he wrote to Fairchild on May 6, 1925. "I was trained under you and I came into this work with the understanding that I was going to run it more or less along the lines you have followed. Now they don't approve of many of your policies. I have constantly struggled, these three years, to keep alive your ideals and your policies for the office. It has been a losing fight."

Popenoe, who had finally fallen in love and married an English botanist named Dorothy Hughes, decided to take her and their infant son to the tropics that he loved. In July 1925 he accepted a job with the United Fruit Company in Honduras and resigned from the agriculture department. Roland McKee, a forage crops expert in the office who Popenoe said did not share Fairchild's ideals, was temporarily put in charge of seed and plant introduction. Soon department officials became impatient with Fairchild's long absences.

"He had always been head of the office, but it was almost a joke around the department for he was never there," said Knowles A. Ryerson. "There was always somebody 'acting.'" Ryerson, a well-known horticulturalist, had been a childhood friend of Wilson Popenoe in Altadena and spent years cultivating plants in California, Haiti, and Palestine.

At the end of 1927 Beverly Galloway invited Ryerson, then thirty-five, to lunch in Washington. After the meal, Galloway, as he had done after a different lunch thirty-eight years earlier, offered Ryerson a job, but this time it was David Fairchild's own job. Ryerson accepted it on the condition that the position would be permanent, not acting, and Galloway agreed. Ryerson took over the office on January 5, 1928. Unlike Fairchild, he adopted an unromantic title: "Principal horticulturalist in charge of the division of foreign plant introduction." Ryerson continued to send employees out to find new material, but the expeditions lacked the romance that Fairchild loved. Gathering genes and chromosomes was the objective, Ryerson told an interviewer many years later.[5]

Fairchild was initially gracious about Ryerson's appointment, even hosting a celebratory lunch for him at the Cosmos Club. Fairchild continued to collect his full salary of $2,800 a year, but he agreed to assume the lesser title of "senior horticulturalist and specialist."

In interviews later about his own accomplishments, Ryerson said he respected Fairchild and his work, even though, unlike the other American plant hunters, his expeditions had been conducted in great comfort.

"When it came to plant exploring, Fairchild's world tour of botanic gardens was luxury first-class, never rough and tumble," Ryerson said. "No, he wasn't doing the Frank Meyer or the Wilson Popenoe or Joe Rock type of thing. But, nevertheless, at that time, his contacts, the observations he made, and the plants and materials he sent back were of great value."[6]

About five years after Ryerson took over, however, as the Great Depression wracked America, the department couldn't afford to pay Fairchild as much money. Budget cutbacks forced the agency to fire seventeen people and stop all foreign travel. Remaining workers took a 20 percent salary cut. Galloway and Dorsett retired, and Swingle saw his plant-breeding work drastically reduced. Ryerson concluded that, while he could not continue to pay Fairchild his full salary, he could justify keeping him on the payroll at half-salary because his work was still valuable to the department. "Some of us are known in particular locations, but his name is known around the world," Ryerson explained.

Ryerson also wanted to protect Fairchild from public embarrassment. A cut in salary was "better for him than to have it come out in the *Congressional Record* that during a depression the son-in-law of Alexander Graham Bell is still on top salary with a winter home in Florida," Ryerson said later.

This development infuriated Fairchild. "It hurt his pride no end," Ryerson said. "He thought he should have full salary for all the things he had done in the past. . . . He *had* built the Office of Plant Introduction, which was a great contribution."

Fairchild accepted a reduction in salary to $1,400 a year only after Ryerson went to Florida to ask Marian to intervene. "She understood the problem and saw that it was the only thing to do," Ryerson said, adding he was grateful for her help. "She was a great lady and a warm, very understanding person."[7]

Fairchild continued to collect the reduced salary until spring 1935 when officials pushed him to take early retirement, a move that freed his $1,400 a year for other plant introduction work. He was reluctant, but he did it and became only a collaborator—an unpaid one because dollar-a-year salaries had been eliminated. "Naturally a man feels it to have Father Time call a halt to his career," he admitted in an April 25, 1935, letter to Frederick D. Richey, a senior department official. "Forty-six years seems a long time to be associated with a government, and to end that association and become a collaborator only looks like a step further towards the end of things."

He had financial concerns, too. Despite her father's success, Marian Fairchild frequently said she and her husband were "hard up" for money, and David Fairchild often described himself as "land poor." In 1937, the family moved into a small cottage at the Kampong so they could rent out the main house to cover their expenses. Yet Fairchild ended his 1935 resignation letter on a gracious note. "All power to you in your work to keep the old ship afloat on these new seas that roll up around us," he wrote.

Fairchild's last official day on the agriculture department's staff was June 30, 1935. As of that date the office he established had introduced 111,857 varieties of seeds and plants to America.

*   *   *

Almost eighty years later, it is impossible to calculate the value of so many plants from so many places. Even Beverly Galloway, who had supervised Fairchild's work, couldn't generalize about their importance. "Many of the immigrants have their little day or hour and are never again heard from," he wrote in the 1928 *Yearbook of Agriculture*. "Others sink out of sight for a time and later achieve great prominence."

He could have added that a few were out-and-out flops and others were impractical curiosities that Fairchild showed off to his friends and relatives. Yet many of David Fairchild's plant immigrants were great successes of incalculable value. Mark Carleton's durum wheat and Frank Meyer's soybeans completely transformed American agriculture in the twentieth century. And by the beginning of the twenty-first century, Walter Swingle's dates and figs and Wilson Popenoe's avocados had become staples of the American diet. Meyer's lemon was a food lovers' delight. Many other introductions served the important but less visible role of providing essential breeding material to make existing plants hardier or more productive.

It was an accomplishment that was eventually recognized by the Department of Agriculture. In 1940 Secretary Henry A. Wallace honored Fairchild in a speech during a ceremony at a government experiment station in Maryland. "The garden of your dreams has been not merely a pretty thing; it is woven into the very fabric of our lives," Wallace said. "Barbour Lathrop's dream and yours and that of the host of friends who have helped you on your way—that dream of taking what is good in the plants of all the earth to make our country better—this has grown as so many of your plants have grown."

*   *   *

When David Fairchild left Washington in 1924, after giving up a job that kept him at a desk for most of twenty years, his weariness suddenly vanished. Overnight, it seems, he acquired enormous energy and enthusiasm that propelled him into a constant series of adventures that filled the rest of his life. "As the fieldmen used to say, DF had it made," Ryerson said later.

His first project took him back to the tropics. While he waited for Allison Armour to outfit his ship for the scientific expedition, Fairchild helped his friends William Morton Wheeler and Thomas Barbour, an entomologist and a zoologist associated with Harvard University, set up a new scientific research center on an island in Panama's rainforest. Initially called the Barro Colorado Island Biological Laboratory (and now known as the Smithsonian Tropical Research Institute), the facility was modeled after the botanical institutes Fairchild loved in Naples and Java. Fairchild spent part of the summer of 1924 building a laboratory on the island. He also persuaded Barbour Lathrop to donate the first sixty books for the institute's library as well as pay for a small launch. As a reward for their support, David Fairchild, Barbour Lathrop, Thomas Barbour, William Wheeler, and even Allison Armour have jungle trails there named in their honor. Fairchild loved setting up the tropical laboratory. "That stay with you was one of the most wonderful times I have had or ever expect to have in my life," he told Wheeler afterwards.

In September 1924, David and Marian Fairchild—and sometimes their children and friends—began exploring for plants, often under Allison Armour's sponsorship. They drove an old American car through Algeria and Morocco, visiting gardens, ancient cities, and souks. They especially enjoyed Mogador, then a drowsy little town on the sea that was home of the rare argan nut trees. Marian Fairchild showed off her firm feminist convictions by driving their Dodge sedan through Fez. "Marian takes every opportunity to run the car around through the narrow streets just to show that she is not in any way under her husband's thumb," Fairchild told Grosvenor on April 4, 1925.

Allison Armour's ship, a converted, 1,315-ton tramp steamer he called the *Utowana*, carried them next through the Mediterranean, the Red Sea, and the Arabian Sea to Ceylon. At the end of 1925, after the *Utowana* suffered engine troubles, the Fairchilds took a commercial steamship to Sumatra.

"I have experienced all kinds of lazy feelings, all kinds of daydream sensations, but I declare that those on a boat in a tropical sea are the most curiously romantic," he wrote later. "Many of the passengers on board would, I believe, have been perfectly willing that the boat should sail straight out into the sunset, leaving the watery sea for the sea of the sky with its cloud islands, and go on and on forever."[8]

Sumatra and nearby islands were full of fascinating, mysterious plants. In April 1926, Fairchild finally took Marian to Java, fulfilling a promise he had made when they married more than twenty years earlier. Soon after they arrived, they visited a penal colony off the coast of Java where they encountered an imprisoned headhunter. He "had failed to get as many heads as his sweetheart demanded before she would marry him," Fairchild explained, "because the government stopped him and sent him here after his last murder." He had only five; she wanted six. In Java Marian discovered she enjoyed being carried in a sedan chair.[9]

A few days later, they discovered a new fruit called the kepel that grew rampant in a bygone sultans' abandoned water palace. The palace had been built in the eighteenth century as a lounge for the sultans' wives and concubines. Once luxuriously appointed, the concrete structure was crumbling when Fairchild discovered it. The pools were clogged with floating water plants, and beautiful pink and red kepel trees framed its doorways.[10]

The kepel, whose proper name is *Stelechocarpus burahol*, is related to the cherimoya and the pawpaw, both fruits Fairchild had promoted in America. Local guides told the Fairchilds that sultans had planted the trees and ordered their lovers to eat kepel fruit because it made their bodily fluids smell like violets. They also warned outsiders that stealing the forbidden fruit would bring bad luck.[11] Fairchild immediately went to the open market in Djokjakarta to buy some for America. (Kepel was the 67,491st seed or plant to arrive in Washington from the ends of the earth. In 2012 the plant was growing at The Kampong in Coconut Grove.)

At the age of fifty-seven, in a beautiful, rundown spot far away from home, Fairchild had discovered one of the world's most romantic fruits. "That's the best we get out of life anyway—romance," he said years later.[12]

* * *

David and Marian Fairchild continued to travel with Allison Armour over the next seven years, visiting the West Coast of Africa, Constantinople, the West Indies (on one trip they stopped at seventy-two ports in twenty-six countries, Fairchild reported), everywhere Fairchild's curiosity and Armour's generosity would take them.

Between trips he joined Marjory Douglas on Ernest F. Coe's early campaign to save the Florida Everglades by becoming the first president of the Tropical Everglades Park Association and writing newspaper articles about the natural glories of the swamp. "The Everglades of South Florida have a strange and to me appealing beauty," he said during a speech on February 28, 1929. "Their charm partakes of the charm of the Pacific Islands." With the authority of a global traveler, he insisted that the Everglades' natural beauty was unmatched anywhere in the world.

Fairchild's adventures with Armour ended in April 1933 when he became gravely ill after contracting an infection in the Caribbean. He recovered slowly and replaced most traveling with writing and speaking, activities that allowed him to constantly relive his adventures. Fairchild's many books and articles brought attention to his accomplishments and led to the establishment of the Fairchild Tropical Botanic Garden in Coral Gables by Colonel Robert H. Montgomery, yet another wealthy philanthropist who loved nature—he collected trees, large ones—and was charmed by David Fairchild.

The project began by accident. One day in 1936 Montgomery, an accountant and business executive with a home in Florida, was playing bridge with Stanton Griffis, a New York investor and businessman. Griffis said he wanted some land near Miami, so Montgomery obligingly bought twenty-five acres for him. But Griffis backed out of the deal, leaving Montgomery with land he didn't need. The situation gave Montgomery the opportunity to create a garden of palms.[13] This palmetum soon expanded into the eighty-three-acre site that is now the Fairchild Tropical Garden. The garden officially opened on March 23, 1938. Griffis became one of its first lifetime members.

Montgomery and Fairchild's love of palm trees led to Fairchild's last big seagoing adventure. His sponsor this time was Anne Archbold, the daughter of an executive of the Standard Oil Company of New Jersey. She arranged to take David and Marian Fairchild to the northern islands of the Moluccas, a remote region once known as the "Spice Islands" which Fairchild had never toured, to collect palm seeds. To get there she hired shipbuilders in Hong Kong to build a hundred-foot replica of a fifteenth-century Chinese junk. She called the boat the *Cheng Ho* in honor of an early Chinese explorer. The Archbold expedition left California in September 1939, carrying scientific equipment and, as always, small gifts for South Sea natives: dime store jewelry, tiny bottles of perfume, and toy balloons.

The Fairchilds' friends were shocked that they would risk another long voyage, especially when the world was on the brink of another world war. "Marian and I felt, however, that precisely because we were older the world could spare us more easily if we never came back from the other side of the globe, and I am sure that Marian, at least, secretly enjoyed doing something that her friends thought dangerous!" Fairchild wrote. "Besides, in our philosophy dangers to health lurk everywhere and the majority of folk seem to die in their own houses."[14]

They survived the trip—Fairchild was able to celebrate his seventy-first birthday in the botanical garden in Buitenzorg—but the outbreak of the Second World War cut the voyage short before Fairchild was realized his dream of exploring the Spice Islands. The junk made only four brief stops in the Moluccas, but they were long enough for Fairchild to add a few new names to the list of remote places he had visited: the Batchan islands, Halmahera, Morotai, and the Talauds.

Despite its brevity, the Archbold Expedition collected many seeds for the Fairchild Tropical Garden, including ninety-nine packages of palm seeds. Because of the work of Alec Bell and his associates many years earlier, Fairchild was able to send them to Florida by air express. The seeds were a welcome addition to the garden, which quickly became a beautiful center dedicated to tropical plants. It has offered programs for research, education and conservation, as well as opportunities to eat Fairchild's favorite foods.

One of the most significant festivals was held eight years after the garden opened. In April 1926, during an early trip to Malaysia sponsored by Allison Armour, Fairchild bought hundreds of mangosteens in the market at Penang and sent the seeds to Wilson Popenoe, who was setting up the Lancetilla Agricultural Experiment Station in Tela, Honduras.

FIGURE 30. David Fairchild selects mangosteens in the Penang market in April 1926. Seeds from these fruit were planted in Honduras and eighteen years later produced forty tons of mangosteens, more than enough for Fairchild to feed guests at a mangosteen party in Florida. Photo courtesy of Hunt Institute for Botanical Documentation, Carnegie Mellon University, Pittsburgh.

Popenoe planted the seeds and waited. Mangosteens are difficult plants to grow for they need the right soil and climate and, most significantly, more time than commercial growers want to give them, especially in America. However, by 1944 the orchard had produced thirty tons of David Fairchild's favorite fruit. The next year officials at Lancetilla flew crates of the special treat to Fairchild, enough for him to host a mangosteen party for three hundred people at the garden. It was August 18, 1945. "This a great moment for me," Fairchild told a writer for the Garden's newsletter. "Fifty years ago I first tasted a mangosteen in a great garden of Buitenzorg in Java and selected it as, for me at least, the most delicate of all fruits. I made up my mind that I would introduce it into the United States. Fifty years is a long time. But this is the day."

Fairchild slowed down after the war, staying home in Florida and continuing to write and speak about his travels. He outlived most of his old friends, but not before awarding most of them the Meyer Medal. He earned one himself in 1939.

When Walter Swingle turned eighty in 1951, Fairchild sent him a tribute to their work together in science and horticulture. "In doing this we have seen, as few others have, this planet," he wrote. "Furthermore, we have seen it shrink to so small a globe that we can fly around it in a few days."[15]

Swingle died a year later, and Fairchild wrote another warm note to his widow, Maude. "I don't believe there were ever two boys who have had more romantic times in the world of science than have had we two," he said.

By the middle of 1954, Fairchild's own health had deteriorated. He died at home in Coconut Grove on the afternoon of August 6, 1954. He was eighty-five.

Marian Fairchild immediately sent a telegram with the news to Wilson Popenoe in Honduras. It read: "Your beloved chief has set out on his last trip."

Acknowledgments

I bow with great respect and gratitude to the legions of unnamed librarians who carefully preserved the scores of obscure books and articles I read for this book. I believe I was the first person to open some of them, even though they had been sitting on library shelves for more than a hundred years. And I am especially grateful to the workers who put a lot of the material online. I could never have done my research without their dedication to saving records for posterity.

Of course, some of the librarians can be named. I particularly want to thank Nancy Korber of the Fairchild Tropical Botanic Garden in Coral Gables, Florida, and Angela Todd at the Hunt Institute for Botanical Documentation in Pittsburgh. They were supportive and always generous in sharing their bounty of documents and photographs.

Special acknowledgment goes to the Eugenics Archives, an online record of documents about the American eugenics movement that is maintained by the Cold Spring Harbor Laboratory in Cold Spring Harbor, New York. The website is http://www.eugenicsarchive.org/eugenics/. Posting the collection helps protect America from repeating the horrors inspired by the movement.

Many people at many other places gave me information. They include J. Dustin Williams and Jamie Shriver of the Hunt Institute; John R. Waggener, Malissa Suek, and Shannon E. Bowen of the American Heritage Center at the University of Wyoming; Ann Parsons, Ann Schmidt, and Annemarie Furlong at the Kampong in Coconut Grove, Florida; Jerry Wilkinson and Ken Stinson of South Florida; Maureen Horn of the Massachusetts Horticultural Society; Sheila Connor, Larissa Glasser, and Lisa

Pearson of the Arnold Arboretum; Emelie George, Lynn Stanko, and Sara Lee of the National Agricultural Library's special collections in Beltsville, Maryland; Ernest Gerard of the Dade County Clerk's Office; Michelle Robbins of American Forests; Susan Halpert of the Houghton Library at Harvard University; the Newberry Library; the Library and Archives at the Royal Botanic Gardens at Kew; the National Archives; Steve Hersh, Marcia Evanson, Caroline Harzewski, and Maria R. Estorino of the Otto G. Richter Library's special collection department at the University of Miami; the Entomological Society of Washington of the Smithsonian Institution; the office of the clerk of the Circuit Court of Cook County, Illinois; Benjamin Hayes of the historian's office at the U.S. House of Representatives in Washington, D.C.; Rachel Sailor of the Special Collection Library at the University of Iowa; the Albert and Shirley Small Special Collection Library at the University of Virginia at Charlottesville; the Chicago History Museum; the Southern Historical Collection, Manuscripts Department, Wilson Library, the University of North Carolina at Chapel Hill; Amanda Langendoerfer and Elaine M. Doak of Pickler Memorial Library at Truman State University; Charles B. Greifenstein of the American Philosophical Society in Philadelphia; Florence Palomo at Conde Nast Publications in New York City; the manuscript collection at the Library of Congress; the Coachella Valley Historical Society in Indio, California; Cheri Smith of the Hesburgh Library at the University of Notre Dame; Lynn Gamma and William M. Russell of the U.S. Air Force Historical Research Agency at Maxwell Air Force Base in Alabama; Timothy Fries at the Hilton M. Briggs Library at South Dakota State University; Ashley M. Morton at National Geographic Creative; and Tad Bennicoff at the Smithsonian Institution Archives.

Also I will be forever thankful for the vast permanent collections in New York City: the New York Public Library, the LuEsther T. Mertz Library at the New York Botanical Garden, the libraries at Columbia University, Bobst Library at New York University, and the American Museum of Natural History.

I also thank Helene Pancoast, David Fairchild's granddaughter, for allowing me to use family photographs, and Mariana Field Hoppin, Barbour Lathrop's great-grandniece, for approving my use of quotations from

Marjory Douglas's book about her relative. Susan Freinkel made many helpful suggestions that improved the manuscript.

The top-notch professionals at the University Press of Florida were a pleasure to work with, especially Sian Hunter, my acquisitions editor. Her colleagues included Nevil Parker, Teal Amthor-Shaffer, Elaine Otto, Rachel Doll, and Stephanye Hunter.

I also thank Susan Rabiner and Sydelle Kramer for helping me turn my project into a book. And I am most grateful for my wonderful friends' unflagging kindness and encouragement: Kathy Jones, Clyde Haberman, Alison Mitchell, Frank Clines, Kit Seelye, Bobbie Grenier, Jean-Louis Grenier, and the late Hethea Nye.

My biggest gratitude goes to Drew Fetherston, my husband and fellow journalist, for his help and support and for making early twentieth-century plant exploring one of the many adventures we have shared.

Notes

Chapter 1. Escape from Kansas

1. D. Wilder, *Annals of Kansas*, 643.
2. L. I. Wilder, *On the Banks of Plum Creek*, 195.
3. Carey, *Kansas State University: The Quest for Identity*, 59–60.
4. Fairchild, *The World Was My Garden*, 3.
5. Ibid., 7.
6. Douglas, *Adventures in a Green World*, 14.
7. Fairchild, *The World Was My Garden*, 229.
8. Wallace, *The Malay Archipelago*, 2:102.
9. Fairchild, *Garden Islands*, vii.
10. Fairchild, "Byron David Halsted, Botanist," 2.
11. Fairchild, *The World Was My Garden*, 16.
12. Galloway to Rock, September 14, 1923, National Archives, RG 54, general correspondence.
13. Fairchild, *The World Was My Garden*, 16.
14. Ibid., 19.
15. Ibid., 27.
16. Ibid., 28.
17. Ibid.
18. Ibid., 28–29.
19. Ibid., 30.
20. Fairchild, "Uncle Barbour."
21. Fairchild, *The World Was My Garden*, 31.
22. Ibid., 32.
23. Fairchild, "Uncle Barbour."

Chapter 2. Good Knight of the Four Winds

1. Fairchild, "Uncle Barbour."
2. Daniel Bryan to Andrew Wylie Jr., May 14, 1861, Wylie family papers.
3. Minna Lathrop to Daniel and Mary Bryan, September 13, 1862, Wylie family papers.

4. Douglas, *Adventures in a Green World*, 4.
5. Ibid., 5.
6. Fairchild, "Panama Trip," 80.
7. Douglas, *Adventures in a Green World*, 5.
8. Lathrop to Daisy Beckwith Lawton, December 4, 1906, Lawton papers, University of North Carolina.
9. Douglas, *Adventures in a Green World*, 6.
10. Cody, *My Life*, 349–50.
11. Annals of the Bohemian Club, 1:25, Field papers at Fairchild Tropical Botanic Garden.
12. Douglas, "My Most Unforgettable Character," 69–70.
13. Renwick, *Recollections*, 47.
14. Douglas, *Adventures in a Green World*, 9.
15. Ibid., 11.
16. Ibid., 13.

Chapter 3. The New Year's Resolution

1. Fairchild, *The World Grows Round My Door*, 10.
2. Fairchild, *Exploring for Plants*, 264.
3. Fairchild, "Uncle Barbour."
4. Douglas, *Adventures in a Green World*, 15.
5. Fairchild, "Uncle Barbour."
6. Fairchild, *The World Was My Garden*, 41.
7. Fairchild, "Our Plant Immigrants," 184.
8. Fairchild, *The World Was My Garden*, 47.
9. Ibid., 49.
10. Fairchild, "Our Plant Immigrants," 184.
11. Fairchild, "Sumatra's West Coast," 450.
12. Fairchild, *Garden Islands*, 4.
13. Thomas Barbour to his parents, January 15, 1907 in *Letters*.
14. Fairchild, *The World Was My Garden*, 71.
15. Fairchild, "Mangosteen, Queen of Tropical Fruits," 1.
16. Ibid.
17. Fairchild, *Exploring for Plants*, 18.
18. Fairchild, *The World Was My Garden*, 80.
19. Fairchild, *Garden Islands*, 9.
20. Douglas, *Adventures in a Green World*, 17.
21. Ibid.
22. Fairchild, "Sumatra's West Coast," 450.
23. Fairchild, *The World Was My Garden*, 82.
24. Fairchild, "New Hope of Farmers," 7731.
25. Fairchild, *The World Was My Garden*, 84.
26. Douglas, *Adventures in a Green World*, 2.

Chapter 4. The Golden Age of Travel

1. Jefferson, *Memoirs, Correspondence, and Private Papers*, 144.
2. Wulf, *The Founding Gardeners*, 77.
3. Wilkinson, *Dr. Henry Edward Perrine*.
4. Account taken from Bellamy, "The Perrines at Indian Key," and Wilkinson, "The Massacre Story."
5. Fairchild, *The World Was My Garden*, 92.
6. Ibid., 86.
7. Ibid., 88.
8. Fairchild, "Uncle Barbour."
9. Fairchild, *The World Was My Garden*, 96.
10. Fairchild, "Uncle Barbour."
11. Fairchild, *The World Was My Garden*, 96, 97.
12. Ibid., 99.
13. Ibid., 104.
14. Wilcox and Wilson, *Tama Jim*, 16.
15. Fairchild, *The World Was My Garden*, 105.
16. Fairchild, *The World Grows Round My Door*, 18.
17. Fairchild, *The World Was My Garden*, 106.
18. Fairchild, "Uncle Barbour."
19. Swingle, "Some Citrus Fruits That Should Be Introduced into Florida," 117.
20. Douglas, "Two Men and a Garden," 15.
21. Fairchild, *The World Was My Garden*, 106.
22. Fairchild, *Systematic Plant Introduction*, 9.
23. Fairchild, "Foreign Plant Introduction Medal," 170.
24. Fairchild, "Uncle Barbour."
25. Fairchild, *The World Was My Garden*, 119.

Chapter 5. Tramps Together

1. Fairchild, *The World Was My Garden*, 123.
2. Ibid., 131.
3. Ibid., 125.
4. Fairchild to Lathrop, November 28, 1924, Correspondence.
5. USDA, inventory of plants no. 8, Plant Inventory no. 3911.
6. Fairchild, *The World Was My Garden*, 141–42.
7. Ibid., 147.
8. Ibid., 148.
9. Ibid.
10. Ibid., 151.
11. Fairchild, "Free From Interruptions!" 564.
12. Knapp, Journal, 81.
13. Fairchild, *The World Was My Garden*, 154.
14. Ibid., 157–58.

15. Ibid., 160.

16. Ibid., 181.

17. June 25, 1940, memo from David Fairchild, in Fairchild correspondence, Montgomery Library, Fairchild Tropical Botanic Garden.

Chapter 6. From Far East to Mideast

1. Lathrop to Fairchild, October 21, 1901, *Green World*, 30.

2. Fairchild, *The World Was My Garden*, 212.

3. Ibid., 219.

4. Ibid., 220.

5. Ibid., 228.

6. Fairchild, "A Date-Leaf Boat of Arabia," 1.

7. Fairchild, *The World Was My Garden*, 229.

8. Ibid., 236.

9. Ibid., 240.

10. Ibid., 249–50.

11. Ibid., 253–54.

12. Ibid., 272.

13. Ibid., 274.

14. Ibid., 277.

15. Fairchild, "Real Pioneers."

16. Fairchild, "Our Plant Immigrants," 187.

17. Fairchild, "Some Plant Introduction Experiences," 56.

18. Fairchild, "Testing New Foods," 23.

19. Fairchild, *The World Grows Round My Door*, 272.

Chapter 7. The Ends of the Earth

1. Loen, *With a Brush and a Muslin Bag*, 100.

2. Ibid., 91.

3. Loen, *The Banebryder*, 20.

4. Ibid., 21, 22.

5. Fairchild, "Uncle Barbour."

6. Cunningham, *Frank N. Meyer*, 163.

7. Dies, *Titans of the Soil*, 145.

8. Fairchild, *The World Was My Garden*, 116.

9. Carleton, "Hard Wheats Winning Their Way," 409.

10. Carleton and Chamberlain, *The Commercial Status of Durum Wheat*, 20.

11. Fairchild, "American Fiction of Polished Rice."

12. Roeding, "Future of the Fig Industry," 207.

13. Roeding, "How the Smyrna Fig and Fig Insect Came to California."

14. Swingle, "Some Points in the History of Caprification," 180.

15. Condit, *The Fig*, 42.

16. Swingle, "Some Points in the History of Caprification," 181.

17. Fairchild, "Walter Tennyson Swingle."
18. Rogers, *Erwin Frink Smith,* 202.
19. Fairchild, "Walter Tennyson Swingle."
20. Venning, "Walter Tennyson Swingle, 1871–1952," 116.
21. Condit, *The Fig,* viii.
22. Venning, "Walter Tennyson Swingle, 1871–1952," 14.
23. Roeding, *Smyrna Fig at Home,* 44.
24. Roeding, "How the Smyrna Fig and Fig Insect Came to America."
25. Venning, "Walter Tennyson Swingle, 1871–1952," 13.
26. Coachella Valley Historical Society, "In the Beginning . . . The Story of Dates, part 1," 23.
27. Ibid., 24.
28. Venning, "Walter Tennyson Swingle, 1871–1952," 19.
29. Letter from Swingle to Charles Lummis, October 22, 1900, http://libraries.theautry.org/2013/03/14/piecing-it-all-together-part-ii-3/.
30. Marian Fairchild to Mabel Bell, June 6, 1910.

Chapter 8. Romance in America

1. Douglas, *Adventures in a Green World,* 20.
2. Wentzel, "Gilbert Hovey Grosvenor, Father of Photojournalism."
3. Fairchild, *The World Was My Garden,* 195.
4. Zuckerman, *The Kampong,* 40.
5. Ibid.
6. Ibid.
7. Mabel Bell to Alec Bell, June 30, 1883, qtd. in Toward, *Mabel Bell,* 119.
8. Casey and Mary Borglum, *Give the Man Room,* 79.
9. Ryerson, *The World Is My Campus,* 95.
10. Fairchild, *The World Was My Garden,* 290.
11. Howells, "Mildred Howells as the Father's Daughter," 21.
12. Fairchild, *The World Is My Garden,* 311.
13. Zuckerman, *The Kampong,* 42.
14. Burger, "Alexander Graham Bell ("Sandy") Fairchild: A Biography," ix.

Chapter 9. Flying Machine Crank

1. Bell, journal entry, July 17, 1901, in Bell family papers.
2. Qtd. in Toward, *Mabel Bell,* 169.
3. "How the Airplane Made Its Public Bow," *New York Times,* March 18, 1928.
4. McCurdy, letter to "George," 1908, Bell family papers.
5. Fairchild, *The World Was My Garden,* 334.
6. "Winning the *Scientific American* Trophy July 4, 1908 by Mrs. David G. Fairchild," association's copy, Bell family papers.
7. Fairchild, *The World Was My Garden,* 343.
8. Curtiss and Post, *The Curtiss Aviation Book,* 54.

9. Quotations from "Airship Falls," *Washington Post*, September 18, 1908.
10. Fairchild, *The World Was My Garden*, 351.

Chapter 10. A Beautiful Job

1. Ernest Wilson, *Aristocrats of the Garden*, 290.
2. Cunningham, *Frank N. Meyer*, 16.
3. Smith, *Frank N. Meyer*, 336.
4. Ibid.
5. Cunningham, *Frank N. Meyer*, 13.
6. Ibid.
7. Fairchild, *The World Was My Garden*, 315.
8. Fairchild, "Explorers and Explorations," 7.
9. Fairchild, "An Agricultural Explorer in China," 7.
10. Cunningham, *Frank N. Meyer*, 35.
11. Ibid., 43.
12. Ibid., 42.
13. Van Uildrinks, "De Aarde en Haa Volken," 9.
14. Cunningham, *Frank N. Meyer*, 44.
15. "Capital's Columbus Starts This Week," *Washington Post*, August 15, 1909.
16. Van Uildrinks, "De Aarde en Haa Volken," 13.
17. Ibid., 11.
18. Cunningham, *Frank N. Meyer*, 63.
19. Ernest Wilson, *Naturalist in Western China*, 1:27.
20. Cunningham, *Frank N. Meyer*, 65.
21. Meyer, *Agricultural Explorations*, 44.
22. Fairchild, *The World Was My Garden*, 443.
23. Meyer to Massachusetts Horticultural Society, March 25, 1916.
24. Ernest Wilson, "Charles Sprague Sargent."
25. Fairchild, *The World Was My Garden*, 381.
26. Ibid., 315.
27. Ibid., 346.

Chapter 11. Easy Money

1. Wilson Popenoe, unpublished autobiography, 3.
2. Hooper, "Rev. Sheldon a Journalist at Heart."
3. Popenoe, unpublished autobiography, 4.
4. Hoffman, "The Man Who Saves Marriages," 123.
5. Wilson Popenoe, unpublished autobiography, 3.
6. F. O. Popenoe, *Tropical Fruits for California*, 17.
7. California Avocado Society, "In Memorium Carl B. Schmidt."
8. Mason, "The Date in California," 472.
9. Wilson Popenoe, unpublished autobiography, 7.
10. Ibid.

11. Paul Popenoe, "Unexpected Commission," 48.

12. Paul Popenoe, *Date Growing*, xiii.

13. Rosengarten, *Wilson Popenoe: Agricultural Explorer*, 19.

14. Wilson Popenoe, "David Fairchild, Plantsman," 70.

15. Dorsett, Shamel, and Popenoe, *The Navel Orange of Bahia*, 6.

16. Galloway, "Historic Orange Tree," 163.

17. Fairchild, "Plant Introduction for the Plant Breeder," 412.

18. National Archives, expedition reports, roll 6, vol. 21.

19. Wilson Popenoe, unpublished autobiography.

20. Fairchild, *The World Was My Garden*, 423.

21. Fairchild, "The Jaboticaba, the Grape of Brazil," 1.

22. Rosengarten, *Wilson Popenoe*, 32.

23. Flipse, "The Avocado from the Investor's Standpoint," 62.

24. Cunningham, *Frank N. Meyer*, 220.

25. Rosengarten, *Wilson Popenoe*, 54.

26. Wilson Popenoe, "Hunting Avocados," 180.

Chapter 12. Better Babies

1. Royal Horticultural Society, "Hybrid Conference Report," 46.

2. Troyer and Stoehr, "Willet M. Hays, Great Benefactor," 435.

3. Webber, "Explanation of Mendel's Law of Hybrids," 138.

4. Speech awarding Meyer Medal to Fairchild, July 24, 1939, USDA news release.

5. American Breeders' Association, *Proceedings*, 1:9.

6. Fairchild, *The World Was My Garden*, 294.

7. Galton, *Inquiries into Human Faculty and Its Developments*, 17.

8. American Breeders' Association, *Proceedings*, 2:11.

9. American Breeders' Association, *Proceedings*, 4:202.

10. Kimmelman, "The American Breeders' Association: Genetics and Eugenics in an Agricultural Context," 171.

11. "Experimental Evolution on Long Island," *New York Times*, June 3, 1906.

12. Davenport, "Eugenics," 69.

13. MacDowell, "Charles Benedict Davenport," 29.

14. Ibid., 3.

15. Ibid., 29.

16. Fairchild, *The World Was My Garden*, 403.

17. Fairchild to Davenport, March 19, 1913, Davenport papers.

18. Qtd. in *Journal of Heredity*, January 1914, 33.

19. Fairchild, *The World Was My Garden*, 424.

20. MacDowell, "Charles Benedict Davenport," 30.

21. Watson, "Genes and Politics," 10.

22. Laughlin, "Calculations," 478.

23. Ibid., 480.

24. Hassencahl, "Harry H. Laughlin," 108.

25. Davenport to Fairchild, June 22, 1922, Davenport papers.

26. Hoffman, "The Man Who Saves Marriages," 123.
27. David Popenoe, *Remembering My Father*.
28. Paul Popenoe, "German Sterilization Law," 260.
29. Paul Popenoe, "Progress of Eugenic Sterilization," 25.
30. Kevles, *In the Name of Eugenics*, 164.

Chapter 13. A Chinese Wall

1. Mallis, *American Entomologists*, 88.
2. Sasscer, "Charles Lester Marlatt," 2.
3. Howard, *Fighting the Insects*, 29.
4. Ibid., 30.
5. Howard, *A History of Applied Entomology*, 93.
6. Galloway, *Proceedings of the National Convention for the Suppression of Insect Pests and Plant Diseases*, 8.
7. Ibid., 9, 11.
8. Marlatt, *Entomologist's Quest*, 38.
9. Ibid., 27.
10. Marlatt, *San Jose Scale*, 169.
11. Ibid., 170.
12. Marlatt, *Entomologist's Quest*, 273.
13. Ibid., 308, 311.
14. "Would Not Have a Physician," *Washington Post*, October 29, 1903, 2.
15. Marlatt, *San Jose Scale*, 155.
16. Bates, *A Jungle in the House*, 44.
17. Howard, *A History of Applied Entomology*, 98.

Chapter 14. Plant Enemies

1. Howard, *A History of Applied Entomology*, 137.
2. Marlatt, "Losses Caused by Imported Tree and Plant Pests," 75.
3. Fairchild, *The World Was My Garden*, 411.
4. Ibid., 415.
5. Fairchild, "Cherry Blossoms of Japan," 1.
6. "Arbor Day Observed," *Evening Star*, March 27, 1908.
7. "Capital's Cherry Blossoms Gift of Japanese Chemist," *Sunday Star*, April 11, 1926.
8. Jefferson and Fusonie, *The Japanese Flowering Cherry Trees of Washington, D.C.*, 10.
9. Fairchild, *The World Was My Garden*, 412–13.
10. Ibid., 413.
11. "Topics of the Times," *New York Times*, January 31, 1910.
12. "Gift Is Destroyed," *Evening Star*, January 29, 1910.
13. Fairchild, *The World Was My Garden*, 425.
14. Cunningham, *Frank N. Meyer*, 227.
15. Marlatt, "Losses Caused by Imported Tree and Plant Pests," 78, 79.

16. Fairchild, "Independence of American Nurseries," 216.
17. Weber, *Plant Quarantine*, 38.
18. Marlatt, "Protecting the United States from Plant Pests," 211.
19. Wiser, *Protecting American Agriculture*, 24.

Chapter 15. The Impossible

1. Cunningham, *Frank N. Meyer*, 110.
2. Ibid., 124.
3. Note on back of photograph (image id: olvwork280702) of *Caragana jubata*, taken March 7, 1911, near Tamgha, Tash, in Arnold Arboretum's visual information access database at http://via.lib.harvard.edu.
4. Cunningham, *Frank N. Meyer*, 127.
5. Ibid., 132.
6. Ibid., 129.
7. Ibid., 137.
8. Ibid., 128.
9. "Cold Resisting Fruits Found," *Los Angeles Times*, May 19, 1912, VII 6.
10. Cunningham, *Frank N. Meyer*, 147.
11. Ibid., 183–84.
12. Farrer, *On the Eaves of the World*, 275–76.
13. Ibid., 278.
14. Ibid.
15. Ibid., 277.
16. Meyer, "China a Fruitful Field," 208.
17. Cunningham, *Frank N. Meyer*, 221.
18. Ibid., 146.
19. Ibid., 199.
20. Ibid., 209.
21. Ibid., 210.
22. Ibid., 215.
23. Van Uildriks, "De Aarde en Haa Volken," 23.
24. Cunningham, *Frank N. Meyer*, 217.
25. Van Uildriks, "De Aarde en Haa Volken," 23.
26. Cunningham, *Frank N. Meyer*, 217.
27. Ibid., 226.
28. Ibid., 225.
29. Van Uildriks, "De Aarde en Haa Volken," 24.
30. Letter in Isabel Shipley Cunningham collection.
31. Details about Meyer's last hours are recounted in the USDA's *South China Explorations*.
32. Letter in Isabel Shipley Cunningham collection.
33. Fairchild, "Agricultural Explorer in China," 7.

Chapter 16. The Last Explorers

1. Wilson Popenoe to parents, February 2, 1919, Popenoe family papers.
2. Rosengarten, *Wilson Popenoe*, 68.
3. Ibid., 67.
4. Ibid., 69.
5. Ibid., 87–88.
6. Wilson Popenoe, "Reflections, Reminiscences, and Observations," 100.
7. Fairchild, "Explorers and Explorations," 26–27.
8. Sutton, *In China's Border Provinces*, 22.
9. Outhier, "George W. P. Hunt," 36.
10. Rock, "Square Plug," 1.
11. Popenoe to family, November 16, 1921, Popenoe family papers.
12. USDA, Plant inventory no. 67.
13. Walravens, *Joseph Franz Rock*, 393.
14. Fairchild, *The World Was My Garden*, 406.
15. Kingdon-Ward, *In the Land of the Blue Poppies*, 5.
16. Wilson Popenoe, "David Fairchild, Plantsman," 79.
17. Bryan, *National Geographic Society*, 197.
18. Fairchild, "Explorers and Explorations," 25.

Chapter 17. Grumpy Old Bachelor Tramp

1. "The Meyer Medal," *Fairchild Tropical Garden Bulletin*, February 1951, 5.
2. Douglas, *Adventures in a Green World*, 40.
3. Ibid., 42.
4. Ibid., 49.
5. Lathrop to Lawton, December 4, 1906, Lawton papers.
6. Starr, *Americans and the California Dream*, 282.
7. Field on Lathrop, 440–42, Field notebooks.
8. Fairchild, "Barbour Lathrop Bamboo Grove," 243
9. Douglas, *Adventures in a Green World*, 58.
10. Douglas, *Voice of the River*, 131.
11. Lathrop to Fairchild, October 27, 1901, in *Green World*, 31.
12. Fairchild, "Uncle Barbour."

Chapter 18. Pushing On

1. Popenoe to parents, January 15, 1918, Popenoe family papers.
2. Fairchild, *The World Was My Garden*, 473.
3. Fairchild to Grosvenor, June 12, 1925, Grosvenor family papers.
4. Fairchild, *The World Was My Garden*, 473.
5. Ryerson, *The World Is My Campus*, 161–64.
6. Ryerson, Stone oral history, 7.
7. Ryerson, *The World Is My Campus*, 164.
8. Fairchild, *Exploring for Plants*, 329.

9. Ibid., 417, 423.
10. Ibid., 430.
11. Schokman, *Plants of the Kampong,* 325.
12. Fairchild, "Little Plant Introduction Garden," 57.
13. Zuckerman, *The Dream Lives On,* 17.
14. Fairchild, *Garden Islands,* 15.
15. Fairchild to Swingle in letter dated "Beginning of a New Year," 1950, Swingle papers, University of Miami.

Bibliography

Archives and Collections

Bell, Alexander Graham. Family papers, 1834–1974. ID no. MSS51268, Manuscript Division, Library of Congress, Washington, D.C. http://www.loc.gov/collection/alexander-graham-bell-papers/about-this-collection/.

Cunningham, Isabel Shipley. Collection on Frank N. Meyer. Special Collections, National Agricultural Library, U.S. Department of Agriculture, Beltsville, Md.

Davenport, Charles B. Papers. American Philosophical Society, Philadelphia.

Fairchild, David. Correspondence. Montgomery Library, Fairchild Tropical Botanic Garden, Coral Gables, Fla.

Field, Henry. Notebooks and papers. Montgomery Library, Fairchild Tropical Botanic Garden, Coral Gables, Fla. (Includes annals of the Bohemian Club and unpublished material on Barbour Lathrop including rough drafts of "Uncle Barbour.")

Grosvenor, Gilbert H. Family papers, 1827–1981. ID no. MSS57240, Manuscript Division, Library of Congress. Washington, D.C. http://hdl.loc.gov/loc.mss/eadmss.ms005006.

Harris, Alice Barton Hill. Diaries. Author's collection.

Knapp, Seaman A. Journal. Archives and Special Collections, Frazar Memorial Library, McNeese State University, Lake Charles, La. http://ereserves.mcneese.edu/depts/archive/FTBooks/knapp.htm.

Lathrop, Barbour. Papers. Hunt Institute for Botanical Documentation, Pittsburgh.

Laughlin, Harry H. Papers. Special Collections, Pickler Memorial Library, Truman State University, Kirksville, Mo.

Lawton, Alexander Robert. Papers no. 415, Southern Historical Collection, Wilson Library, University of North Carolina at Chapel Hill.

Meyer, Frank N. Papers. Special Collections, National Agricultural Library, U.S. Department of Agriculture, Beltsville, Md.

Meyer, Frank Nicholas, and David Fairchild. South China Explorations: Typescript, July 25, 1916–September 21, 1918. Special Collections, National Agricultural Li-

brary, U.S. Department of Agriculture, Beltsville, Md. http://specialcollections.
nal.usda.gov/Meyer-Exhibit#typescript.

Page, Thomas Nelson. Page Collection. Special Collections, Alderman Library, University of Virginia, Charlottesville.

Popenoe, Paul Bowman. Papers, 1874–1991. Collection no. 04681, American Heritage Center, University of Wyoming.

Popenoe, Wilson. Autobiographical manuscript, family correspondence. Hunt Institute for Botanical Documentation, Pittsburgh.

Rock, Joseph. Papers. Arnold Arboretum, Harvard University, Jamaica Plain, Mass.

Ryerson, Knowles A. Records. Hunt Institute for Botanical Documentation, Pittsburgh.

Swingle, Walter Tennyson. Papers. Archives and Special Collections, Otto G. Richter Library, University of Miami, Coral Gables, Fla.

U.S. Department of Agriculture, Bureau of Plant Industry. Soils and Agricultural Engineering records. Microfilm publication M840, RG 54, National Archives at College Park, Md. (Includes general correspondence, 1900–1940, historical file, 1903–39, Frank N. Meyer records, 1902–18, and expedition reports of the office of foreign seed and plant introductions.)

Wylie, Andrew, Jr. Family papers. Wylie House Museum, Indiana University, Bloomington.

Other Sources

Allen, Garland E. "The Eugenics Record Office at Cold Spring Harbor, 1910–1940: An Essay in Institutional History." *Osiris* 2 (1986): 225–64.

———. *Thomas Hunt Morgan: The Man and His Science.* Princeton, N.J.: Princeton University Press, 1978.

American Breeders' Association. *Proceedings of the First Meeting of the American Breeders Association, Held at St. Louis, Mo., December 29 and 30, 1903.* Annual report, vol. 1. Washington, D.C.: American Breeders' Association, 1905.

———. *Proceedings of the Meeting of the American Breeders' Association Held at Lincoln, Nebraska, January 17, 18, and 19, 1906.* Annual report, vol. 2. Washington, D.C.: American Breeders' Association, 1906.

———. *Proceedings of the Meeting of the American Breeders' Association, Held at Washington, D.C., January 28–30, 1908.* Annual report, vol. 4. Washington, D.C.: American Breeders' Association, 1908.

"The Avocado." *Tropical Fruits for California.* Altadena, Calif.: West India Garden, 1913.

Ball, Carleton R. "The History of American Wheat Improvement." *Agricultural History* 4, no. 2 (April 1930): 48–71.

Barbour, Thomas. *Naturalist at Large.* Boston: Little, Brown, 1943.

Barbour, Thomas, and Rosamond Barbour. *Letters Written While on a Collecting Trip in the East Indies.* Paterson, N.J.: privately printed, 1913.

Bartlett, Harley Harris. "Walter Tennyson Swingle: Botanist and Exponent of Chinese Civilization." *Asa Gray Bulletin* 1, no. 2 (April 1952): 107–32.

Bates, Marston. *A Jungle in the House*. New York: Walker, 1970.

Bell, Alexander Graham. "Aerial Locomotion with a Few Notes on Progress in the Construction of an Aerodrome." *Proceedings of Washington Academy of Sciences* 8 (1907): 407–48.

Bellamy, Jeanne. "The Perrines at Indian Key, Florida, 1838–1840." *Tequesta: The Journal of the Historical Association of Southern Florida* 7 (1947): 69–78.

"Bureaucratic Usurpation under Quarantine." *World's Work* 37 (July 1922): 243–44.

Bourke, John Gregory. *Diaries*. Vol. 2, edited and annotated by Charles M. Robinson. Denton: University of North Texas Press, 2003.

———. *On the Border with Crook*. New York: Charles Scribner's Sons, 1891.

Bruce, Robert V. *Bell: Alexander Graham Bell and the Conquest of Solitude*. Boston: Little, Brown, 1973.

Bryan, C.D.B. *The National Geographic Society: 100 Years of Adventure and Discovery*. New York: Harry N. Abrams, 1987.

Burger, John F. "Alexander Graham Bell ("Sandy") Fairchild: A Biography." *Contributions to the Knowledge of Diptera*. Gainesville: Associated Publishers, 1999.

Burrage, A. C. "The Case against Quarantine No. 37." *Garden Magazine* 32, no. 1 (September 1920): 24–27.

California Avocado Society. "In Memoriam Carl B. Schmidt, February 1892–November 1970." *1970–71 Yearbook of the California Avocado Society*, vol. 54. http://www.avocadosource.com/CAS_Yearbooks/CAS_54_1970/CAS_1970-71_PG_006-007.pdf.

Carey, James C. *Kansas State University: The Quest for Identity*. Lawrence, Kans.: Regents Press, 1977.

Carleton, Mark Alfred. "Hard Wheats Winning Their Way." U.S. Department of Agriculture, *Yearbook of Agriculture*, 391–420. Washington, D.C.: Government Printing Office, 1914.

———. *Macaroni Wheats*. U.S. Department of Agriculture, Bureau of Plant Industry, bulletin no. 3. Washington, D.C.: Government Printing Office, 1901.

———. *Russian Cereals Adapted for Cultivation in the United States*. U.S. Department of Agriculture, Division of Botany, bulletin no. 23. Washington, D.C.: Government Printing Office, 1899.

Carleton, Mark Alfred, and Joseph S. Chamberlain. *The Commercial Status of Durum Wheat*. U.S. Department of Agriculture, Bureau of Plant Industry, bulletin no. 70. Washington, D.C.: Government Printing Office, 1904.

Carlson, John W. C. "A Couple of Old Landmarkers." Speech delivered to the Chicago Literary Club on November 21, 1999.

Casey, Robert J., and Mary Borglum. *Give the Man Room: The Story of Gutzon Borglum*. Indianapolis: Bobbs-Merrill, 1952.

Chock, Alvin K. "J. F. Rock, 1884–1962." *Taxon* 12, no. 3 (April 1963): 89–102.

Coachella Valley Historical Society. "In the Beginning . . . The Story of Dates, part 1." *2006 Periscope*. Indio, Calif., 2006.

Coates, Peter. *American Perceptions of Immigrant and Invasive Species*. Berkeley: University of California Press, 2007.

Cody, Wild Bill. *My Life; or, Life and Adventures of Buffalo Bill, Colonel William F. Cody*. New York: Wiley, 1927.

Condit, Ira J. *The Fig*. Foreword by Walter T. Swingle. Waltham, Mass.: Chronica Botanica, 1947.

———. "Fig History in the New World." *Agricultural History* 31, no. 2 (April 1957): 19–24.

Conroy, Sarah Booth. "She Painted the Town Pink." *Washington Post*, February 1, 1999, A1.

Cooke, Kathy J. "The Limits of Heredity: Nature and Nurture in American Eugenics before 1915." *Journal of the History of Biology* 31 (1998): 263–78.

Cory, Ernest N., W. Doyle Reed, and E. Ralph Sasscer. "Charles Lester Marlatt." *Proceedings of the Entomological Society of Washington* 57 (1955): 37–43.

Cox, E.H.M. *Plant-Hunting in China*. London: Collins, 1945.

Cunningham, Isabel Shipley. "Frank Meyer, Agricultural Explorer." *Arnoldia* 44, no. 3 (Summer 1984): 3–26.

———. *Frank N. Meyer: Plant Hunter in Asia*. Ames: Iowa State University Press, 1984.

Curtiss, Glenn Hammond, and Augustus Post. *The Curtiss Aviation Book*. New York: Frederick A. Stokes, 1912.

Davenport, Charles B. "Eugenics, a Subject for Investigation Rather than Instruction." *American Breeders Magazine* 1, no. 1 (January 1910): 68–69.

Derksen, Leo. "The Restless Frank Meyer." *Panorama* 44, no. 20 (1957): 4–5. In the Isabel Shipley Cunningham Collection on Frank N. Meyer, box 3, folder 104.

Dies, Edward Jerome. *Titans of the Soil: Great Builders of Agriculture*. Chapel Hill: University of North Carolina Press, 1949.

Domhoff, G. William. *The Bohemian Grove and Other Retreats*. New York: Harper & Row, 1974.

Dorsett, P. H., A. D. Shamel, and Wilson Popenoe. *The Navel Orange of Bahia; with Notes on Some Little-Known Brazilian Fruits*. U.S. Department of Agriculture, bulletin 445. Washington, D.C.: Government Printing Office, 1917.

Douglas, Marjory Stoneham. *Adventures in a Green World: The Story of David Fairchild and Barbour Lathrop*. Miami: Field Research Projects, 1973.

———. "My Most Unforgettable Character." *Reader's Digest*, November 1948, 67–71.

———. "Two Men and a Garden." *Fairchild Tropical Garden Bulletin*, January 1978, 10–15.

———. *Voice of the River: An Autobiography with John Rothchild*. Sarasota, Fla.: Pineapple Press, 1987.

Economy, Peter, and Patricia Ortlieb. *Creating an Orange Utopia: Eliza Lovell Tibbets and the Birth of California's Citrus Industry*. West Chester, Pa.: Swedenborg Foundation, 2011.

Eisen, Gustav. *The Fig: Its History, Culture, and Curing*. U.S. Department of Agriculture, Division of Pomology. Washington, D.C.: Government Printing Office, 1901.

Fairchild, David. "An Agricultural Explorer in China." *Asia* 21 (December 1920): 7–13.

———. "Alexander Graham Bell: Some Characters of His Greatness." *Journal of Heredity* 13, no. 5 (May 1922): 195–200.

———. "The American Fiction of Polished Rice: A Food Fad That Is Both Costly and Useless." *Saturday Evening Post*, July 28, 1906, 19.

———. "Another Debt to the Missionaries." *Baptist Missionary Magazine*, February 1908, 71.

———. "The Barbour Lathrop Bamboo Grove." *Journal of Heredity* 10, no. 6 (June 1919): 243–49.

———. "Barro Colorado Island Laboratory." *Journal of Heredity* 15, no. 3 (March 1924): 98–112.

———. "Byron David Halsted, Botanist (1852–1918)." *Phytopathology* 9, no. 1 (January 1919): 1–6.

———. "The Cherry Blossoms of Japan in Washington and the American Dogwoods in Tokyo." Undated, unpublished manuscript. Grosvenor family papers, Manuscript Division, Library of Congress.

———. "A Child's Garden of New Plants." *Youth's Companion* 87, no. 11 (March 13, 1913): 135.

———. "A Coming Fruit: The Mango." *Country Life in America*, February 1907, 426–28.

———. "A Date-Leaf Boat of Arabia." *Botanical Gazette* 34, no. 6 (December 1902): 451–53.

———. "The Discovery of the Chestnut Bark Disease in China." *Science* 38, no. 974 (August 29, 1913): 297–99.

———. "Diseases of the Grape in Western New York." *Garden and Forest* 4, no. 154 (February 4, 1891): 59–60.

———. "The Dramatic Careers of Two Plantsmen." *Journal of Heredity* 10, no. 6 (June 1919): 276–80.

———. "Early Experiences with the Chayote." *Proceedings of the Florida State Horticultural Society* 60 (1947): 172–78.

———. "Explorers and Explorations (Agricultural)." Manuscript, Montgomery Library, Special Collections. Fairchild Tropical Botanic Garden, Coral Gables, Fla.

———. *Exploring for Plants.* New York: Macmillan, 1931.

———. "The Farm: A Home and a Business." *Youth's Companion* 83, no. 18 (May 6, 1909): 219.

———. "Fashions in Foods." *Youth's Companion* 93, no. 36 (September 4, 1919): 471.

———. "First 'Bad Habit'—Sucking Its Thumb." *Journal of Heredity* 10, no. 8 (November 1919): 370–71.

———. "Food Prejudices and Plant Introduction." *Proceedings of the Florida State Horticultural Society* 41 (1928): 12–23.

———. "Foreign Farm Life." *Youth's Companion* 84, no. 10 (March 10, 1910): 127.

———. "Foreign Plant Introduction Medal." *Journal of Heredity* 11, no. 4 (April 1920): 169–73.

———. "Free from Interruptions!" *Harvard Graduates Magazine*, June 1924, 562–67.

———. *Garden Islands of the Great East.* New York: Charles Scribner's Sons, 1948.

———. "The Garden of the Oudaias." *Garden Club of America Bulletin* 4, no. 19 (1932): 17–21.

———. "A Genetic Portrait Chart." *Journal of Heredity* 12, no. 5 (May 1921): 213–23.

———. "George Kennan." *Journal of Heredity* 15, no. 10 (October 1924): 402–6.

———. "The Government Policy of Plant Introduction." In *Transactions of the Massachusetts Horticultural Society for the Year 1912*. Part 1, 35–44. Boston: Massachusetts Horticultural Society, 1912.

———. "Green Leaf in Cherry Blossom." *Journal of Heredity* 6, no. 6 (June 1915): 262.

———. "Hints on How to Travel Successfully, for the Benefit of International Travelers of the Department of Agriculture." Alexander Wetmore papers, Smithsonian Institution Archives, Washington, D.C.

———. "Horse-Radish Culture in Bohemia." *Southern Planter* 61, no. 3 (March 1900): 150.

———. *How to Send Living Plant Material to America*. U.S. Department of Agriculture, Bureau of Plant Industry, Office of Foreign Seed and Plant Introduction. Washington, D.C.: Government Printing Office, 1913.

———. "A Hunter of Plants." *National Geographic* 36, no. 1 (July 1919): 57–77.

———. "Hunting for Plants in the Canary Islands." *National Geographic* 57, no. 5 (May 1930): 607–52.

———. "I Never Tasted It." *Atlantic Monthly* 174, no. 11 (November 1944): 91–95.

———. "The Independence of American Nurseries." *American Forestry* 23, no. 280 (April 1917): 213–16.

———. "An Insect That Looks Like a Leaf." *Volta Review* 12 (1910): 347.

———. "The Jaboticaba, the Grape of Brazil." *Florida Plant Immigrants*. Occasional Paper no. 2. Fairchild Tropical Botanic Garden, January 1, 1939.

———. "A Jungle Botanic Garden: The Sibolangit Garden in the Highlands of Sumatra." *Journal of Heredity* 19, no. 4 (April 1928): 145–58.

———. "The Jungles of Panama." *National Geographic* 41, no. 2 (February 1922): 131–45.

———. "The Little Plant Introduction Garden on Brickell Avenue." *Proceedings of the Florida State Horticultural Society* 50 (1937): 55–59.

———. "Madeira, on the Way to Italy." *National Geographic* 18, no. 12 (December 1907): 751–71.

———. "Making Oil from Cattle-Feed." *Volta Review* 12 (1910): 566.

———. "The Mangosteen, Queen of Tropical Fruits." Undated, unpublished article. Montgomery Library, Fairchild Tropical Botanic Garden, Coral Gables, Fla.

———. "The Mangosteen." *Journal of Heredity* 6, no. 8 (August 1915): 339–47.

———. "The Meyer Medal." *Fairchild Tropical Garden Bulletin* 6, no. 5 (February 1951): 4-8.

———. "The New Hope of Farmers." *World's Work* 12 (July 1906): 7731–33.

———. "New Plant Immigrants." *National Geographic* 22, no. 10 (October 1911): 879–907.

———. "New Plants for Breeders." *American Breeders Magazine* 4, no. 2 (April 1913): 103–12.

———. "Notes of Travel. I: Venezuela." *Botanical Gazette* 28, no. 2 (August 1899): 122–26.

———. "Notes of Travel. II: Payta and the Desert Region of Peru." *Botanical Gazette* 28, no. 3 (September 1899): 203–7.

———. "Notes of Travel. III: Rio and Petropolis, Brazil." *Botanical Gazette* 30, no. 2 (August 1900): 125–30.

———. "Notes of Travel. IV: Coffee Growing in Brazil and the Giant Jequitiba Tree." *Botanical Gazette* 31, no. 5 (May 1901): 352–54. (No. 5 of this series was never published.)

———. "Notes of Travel. VI: The Botanical Institute of Netherlands India." *Botanical Gazette* 31, no. 6 (June 1901): 423–25.

———. "Notes of Travel. VII: A Tropical Forest in Ceram." *Botanical Gazette* 32, no. 3 (September 1901): 218–21.

———. "Notes of Travel. VIII: American Autumn Foliage in Europe." *Botanical Gazette* 33, no. 6 (June 1902): 461–62.

———. "Notes on Conditions in China." *Letters on Agriculture in the West Indies, Spain, and the Orient.* U.S. Department of Agriculture, Bureau of Plant Industry, bulletin no. 27, 22–30. Washington, D.C.: Government Printing Office, 1902.

———. "The Ornamental Value of Cherry Blossom Trees." *Art and Progress* 2 (June 1911): 225–26.

———. "The Philosophy of Plant Introduction." Undated, unpublished article. Montgomery Library, Fairchild Tropical Botanic Garden, Coral Gables, Fla.

———. "Our Plant Immigrants." *National Geographic* 17, no. 4 (April 1906): 179–201.

———. "The Palate of Civilized Man and Its Influence on Agriculture." *Journal of the Franklin Institute* 185, no. 3 (March 1918): 299–316.

———. "Panama Trip, August 17 to September 29th, 1921." *Expedition Reports of the Office of Foreign Seed and Plant Introduction of the Department of Agriculture, 1900–1938.* Records of the Bureau of Plant Industry, Soils and Agricultural Engineering, RG 54, microfilm roll 24-87, National Archives at College Park, Md.

———. *Persian Gulf Dates and Their Introduction to America.* U.S. Department of Agriculture, Bureau of Plant Introduction, bulletin no. 54. Washington, D.C.: Government Printing Office, 1903.

———. "Personal Recollections of George B. Cellon, Horticultural Pioneer of South Florida." *Proceedings of the Florida State Horticultural Society* 58 (1945): 205–9.

———. "Plant and Animal Introduction." *American Breeders' Association Magazine* 1, no. 1 (January 1905): 92–100.

———. "Plant Introduction for the Plant Breeder." U.S. Department of Agriculture, *Yearbook of Agriculture,* 411–22. Washington, D.C.: Government Printing Office, 1911.

———. "Plant Introduction Notes from South Africa." U.S. Department of Agriculture, Bureau of Plant Industry, bulletin no. 27, 13–22. Washington, D.C.: Government Printing Office, 1903.

———. "Plant Introduction Opportunities Open to All the Americas." *Proceedings of the Second Pan American Scientific Congress,* 503–10. Washington, D.C.: Government Printing Office, 1917.

———. Preface to *The Plant World in Florida: From the Published Manuscripts of Henry Nehrling.* Collected and edited by Alfred and Elizabeth Kay. New York: Macmillan, 1933.

————. "Present Condition and Opportunity of the American Genetic Association." *Journal of Heredity* 10, no. 2 (February 1919): 65–67.

————. "Protection for the Plant Breeders." Unpublished article. Arnold Arboretum of Harvard University, Cambridge, Mass.

————. "The Real Pioneers." *Youth's Companion* 85, no. 17 (April 27, 1911): 215.

————. "Reminiscences of Early Avocado Introductions." *California Avocado Association Yearbook* 24 (1939): 44–46.

————. "Reminiscences of Early Plant Introduction Work in South Florida." *Proceedings of the Florida State Horticultural Society* 51 (1938): 11–33.

————. "Report of the Committee on Animal and Plant Introduction." *American Breeders Association Magazine* 4, no. 1 (January 1908): 301–4.

————. "Report of the Committee on Plant and Animal Introduction." *American Breeders' Association Magazine* 5, no. 1 (January 1909): 217.

————. "Rosa Hugonis: A New Hardy, Yellow Rose from China." *Journal of Heredity* 6, no. 6 (June 1915): 429–32.

————. "Saragola Wheat." *Miscellaneous Papers.* U.S. Department of Agriculture, Bureau of Plant Industry, bulletin no. 25, 9–12. Washington, D.C.: Government Printing Office, 1903.

————. "The Sensitive Plant as a Weed in the Tropics." *Botanical Gazette* 34, no. 3 (September 1902): 228–30.

————. "Some Plant Introduction Experiences." *Proceedings of the Florida State Horticultural Society* 44 (1931): 54–58.

————. "A Spadeful of Earth." *Youth's Companion* 91, no. 38 (September 20, 1917): 520.

————. "Spring Morning in a South Florida Garden." *Garden Club of America Bulletin* 4, no. 18 (1931): 13–22.

————. "Strange Creatures Hard to See." *Youth's Companion* 90, no. 3 (January 20, 1916): 31.

————. "Sumatra's West Coast." *National Geographic* 9, no. 11 (November 1898): 449–64.

————. *Systematic Plant Introduction: Its Purposes and Methods.* U.S. Department of Agriculture, Division of Forestry, bulletin no. 21. Washington, D.C.: Government Printing Office, 1898.

————. "Testing New Foods." *Journal of Heredity* 10, no. 1 (January 1919): 17–28.

————. "To Brazil for New Plants." *American Breeders Magazine* 4, no. 3 (July 1913): 175–76.

————. "Travels in Arabia and along the Persian Gulf." *National Geographic* 15, no. 4 (April 1904): 139–51.

————. "Twins: Their Importance as Furnishing Evidence of the Limitations of Environment." *Journal of Heredity* 10, no. 9 (December 1919): 387–98.

————. "Two Expeditions after Living Plants." *Scientific Monthly* 26, no. 2 (February 1928): 97–127.

————. "Two Relatives of the Avocado and Their Reintroduction into Florida." *Proceedings of the Krome Memorial Institute published in Proceedings of the Florida State Horticultural Society* 58 (1945): 170–75.

———. "Two Unknown Modern Languages." *Outlook*, September 4, 1909, 93.

———. "Uncle Barbour." Manuscript, 1934. Montgomery Library, Fairchild Tropical Botanic Garden, Coral Gables, Fla.

———. "Visible Records of Heredity." *Journal of Heredity* 12, no. 4 (April 1921): 174–76.

———. "Walter Tennyson Swingle." *Fairchild Tropical Garden Bulletin* 7, no. 7 (April 1952): 7-8.

———. *The World Grows Round My Door.* New York: Charles Scribner's Sons, 1947.

———. "A World Out of Sight." *Youth's Companion* 86, no. 20 (May 16, 1912): 259.

———. *The World Was My Garden: Travels of a Plant Explorer.* New York: Charles Scribner's Sons, 1938.

Farrer, Reginald. "The Last Ascent of Thundercrown." *Gardeners' Chronicle*, July 3, 1915.

———. "My Second Year's Journey on the Tibetan Border of Kansu." *Geographical Journal* 51, no. 6 (June 1918): 341–54.

———. *On the Eaves of the World.* London: E. Arnold, 1917.

Flipse, L. F. "The Avocado from the Investor's Standpoint." *Proceedings of the Florida State Horticultural Society* 34 (1921): 60–63.

Freinkel, Susan. *American Chestnut: The Life, Death, and Rebirth of a Perfect Tree.* Berkeley: University of California Press, 2009.

Frost, Melvin J., and B. Ira Judd. "Plant Explorer of the Americas." *Economic Botany* 24, no. 4 (October–December 1970): 471–78.

Funigiello, Philip J. *Florence Lathrop Page: A Biography.* Charlottesville: University of Virginia Press, 1994.

Galloway, Beverly T. "An Historic Orange Tree." *Journal of Heredity* 13, no. 4 (April 1922): 163–66.

———, ed. *Proceedings of the National Convention for the Suppression of Insect Pests and Plant Diseases by Legislation Held at Washington, D.C., March 5 and 6, 1897.* Washington, D.C.: Government Printing Office, 1897.

———. "Protecting American Crop Plants against Alien Enemies." Lecture delivered before the Massachusetts Horticultural Society, February 15, 1919.

———. *The Search in Foreign Countries for Blight-Resistant Chestnuts and Related Tree Crops.* U.S. Department of Agriculture, circular 383. Washington, D.C.: Government Printing Office, 1926.

Galton, Frances. *Inquiries into Human Faculty and Its Developments.* London: Macmillan, 1883.

Garden Magazine. "Quarantine 37." *Garden Magazine* 29, no. 4 (May 1919): 155, 157.

———. "Something Wrong in the Works: Who and What Are Really behind Quarantine 37?" *Garden Magazine* 29, no. 3 (April 1919): 109–10.

Hamblin, Stephen F. "Plants and Policies." *Atlantic Monthly* 135, no. 3 (March 1925): 353–62.

Hargreaves, Mary W. M. "The Durum Wheat Controversy." *Agricultural History* 42, no. 3 (July 1968): 211–29.

Hassencahl, Frances Janet. "Harry H. Laughlin: Expert Eugenics Agent for the

House Committee on Immigration and Naturalization, 1921–1931." PhD diss., Case Western Reserve University, Cleveland, Ohio, 1970.

Hodge, W. H., and C. O. Erlanson. "Federal Plant Introduction: A Review." *Economic Botany* 10, no. 4 (October–December 1956): 299–334.

Hoffman, Betty Hannah. "The Man Who Saves Marriages." *Ladies Home Journal*, September 1960.

Hoing, Willard L. "James Wilson as Secretary of Agriculture, 1897–1913." PhD diss., University of Wisconsin, 1964.

Hooper, Heather. "Rev. Sheldon a Journalist at Heart." *Topeka Capital-Journal*, March 12, 2000.

Howard, L. O. "Danger of Importing Insect Pests." U.S. Department of Agriculture, *Yearbook of Agriculture*, 529–52. Washington, D.C.: Government Printing Office, 1897.

———. "A Distinguished Entomologist: A Sketch of Professor Riley." *Farmer's Magazine*, January–February 1890.

———. "Economic Loss to the People of the United States through Insects That Carry Disease." *National Geographic* 16, no. 2 (August 1909): 735–49.

———. *Fighting the Insects.* New York: Macmillan, 1933.

———. *A History of Applied Entomology (Somewhat Anecdotal).* Washington: Smithsonian Institution, 1930.

———. "Smyrna Fig Culture in the United States." U.S. Department of Agriculture, *Yearbook of Agriculture,* 79–106. Washington, D.C.: Government Printing Office, 1901.

Howells, Polly H. "Mildred Howells as the Father's Daughter: Living within His Lines." *Harvard Library Bulletin* 5, no. 1 (1994): 9–28.

Humphrey, Harry Baker. *Makers of North American Botany.* New York: Ronald Press, 1961.

Hyland, Howard L. "History of U.S. Plant Introduction." *Environmental Review* 2, no. 4 (1977): 26–33.

Irwin, J.N.H. "The Veteran Guard of the Bohemian Club." *Overland Monthly* 45, no. 4 (April 1905): 261–71.

Isern, Thomas D. "Wheat Explorer the World Over: Mark Carleton of Kansas." *Kansas History: A Journal of the Central Plains* 23, nos. 1–2 (Spring–Summer 2000): 12–25.

Jefferson, Ronald M., and Alan E. Fusonie. *The Japanese Flowering Cherry Trees of Washington, D.C.* Washington, D.C.: Agricultural Research Service, 1977.

Jefferson, Thomas. *Memoirs, Correspondence, and Private Papers.* Vol. 1. London: H. Colburn and R. Bentley, 1829.

K & W Farms Inc. *History of the Fig.* http://figs4fun.com/Links/FigLink041.pdf.

Kay, Elizabeth D. "David Fairchild: A Recollection." *Huntia* 1 (April 1964): 71–78.

Kevles, Daniel J. *In the Name of Eugenics: Genetics and the Uses of Human Heredity.* Cambridge: Harvard University Press, 1985.

Kimmelman, Barbara A. "The American Breeders' Association: Genetics and Eugenics in an Agricultural Context." *Social Studies of Science* 13, no. 2 (May 1983): 163–204.

King, Charles. *Campaigning with Crook*. New York: Harper and Brothers, 1890.

King, James T. *War Eagle: A Life of General Eugene A. Carr*. Lincoln: University of Nebraska Press, 1963.

Kingdon-Ward, Frank. *In the Land of the Blue Poppies: Collected Plant-Hunting Writings*. Edited by Tom Christopher. New York: Modern Library, 2003.

Kirkwood, William P. "Hansen, America's First Plant Explorer." *American Review of Reviews* 48, no. 4 (October 1913): 443–48.

———. "The North Pole of Alfalfa." *Outlook*, May 28, 1910, 187–96.

———. "The Romantic Story of a Scientist." *World's Work* 17 (April 1908): 10109–20.

Klose, Nelson. *America's Crop Heritage*. Ames: Iowa State College Press, 1950.

———. "Dr. Henry Perrine, Tropical Plant Enthusiast." *Florida Historical Quarterly* 27, no. 2 (October 1948): 189–201.

Kohler, Sue A. *Sixteenth Street Architecture*. Vol. 2. Washington, D.C.: Commission of Fine Arts, 1978.

Kruckeberg, Henry W. *George Christian Roeding*. Sacramento: California Association of Nurserymen, 1930.

Kuhl, Stefan. *The Nazi Connection: Eugenics, American Racism, and German National Socialism*. New York: Oxford University Press, 1994.

Ladd-Taylor, Molly. "Eugenics, Sterilization, and Modern Marriage in the USA: The Strange Career of Paul Popenoe." *Gender* and *History* 13, no. 2 (August 2001): 298–327.

Lathrop, Barbour. "A 'Personal' and What Came of It." *Scribner's Magazine*, January 1880, 453–58.

Laughlin, H. H. "An Account of the Work of the Eugenics Record Office." *American Breeders Magazine* 3, no. 2 (1912): 119–23.

———. *Bulletin No. 10A: Report of the Committee to Study and to Report on the Best Practical Means of Cutting Off the Defective Germ-Plasm in the American Population*. Cold Spring Harbor, N.Y.: Eugenics Record Office, 1914.

———. "Calculations on the Working Out of a Proposed Program of Sterilization." *Proceedings of the First National Conference on Race Betterment*. Held January 8–12, 1914, Battle Creek, Mich. Race Betterment Foundation, 1914.

Laut, Agnes C. "Harvesting the Wheat IV—The New Spirit of the Farm." *Outing Magazine* 52, no. 1 (October 1908): 1-16.

Lawrence, George H. M. "A Bibliography of the Writings of David Fairchild." *Huntia* 1 (April 1964): 79–102.

———. *Plant Explorers and Their Introductions with a Bibliography*. Pittsburgh: Carnegie Mellon University, 1970.

Lee, David W., ed. *The World as Garden: The Life and Writings of David Fairchild*. West Charleston, S.C.: Createspace, 2013.

Lewis, James G. "Theodore Roosevelt's Cautionary Tale." *Forest History Today*, Spring/Fall 2005, 53–55.

Lockwood, Jeffrey A. *Locust: The Devastating Rise and Mysterious Disappearance of the Insect That Shaped the American Frontier*. New York: Basic Books, 2004.

Loen, Helen Hansen. *With a Brush and a Muslin Bag: The Life of Niels Ebbesen Hansen*. Kalamazoo, Mich.: Helen Loen, 2003.

———, ed. *The Banebryder: The Travel Records of Niels Ebbesen Hansen*. Kalamazoo, Mich.: Helen Loen, 2004.

MacDowell, E. Carlton. "Charles Benedict Davenport, 1866–1944: A Study of Conflicting Influences." *Bios* 17, no. 1 (March 1946): 2–50.

Mallis, Arnold. *American Entomologists*. New Brunswick, N.J.: Rutgers University Press, 1971.

Marlatt, C. L. "The Annual Loss Occasioned by Destructive Insects in the United States." U.S. Department of Agriculture, *Yearbook of Agriculture*, 461–74. Washington, D.C.: Government Printing Office, 1904.

———. "The Laisser-Faire Philosophy to the Insect Problem." *Proceedings of the 11th Annual Meeting of the Association of Economic Entomologists*. Washington, D.C.: Government Printing Office, 1899.

———. "Losses Caused by Imported Tree and Plant Pests." *American Forestry* 23, no. 278 (February 1917): 75–80.

———. "Need of National Control of Imported Nursery Stock." *Journal of Economic Entomology* 4, no. 1 (February 1911): 107–26.

———. *New Species of Diaspine Scale Insects: Papers on Coccidae or Scale Insects*. U.S. Department of Agriculture, Bureau of Entomology. Washington, D.C.: Government Printing Office, 1908.

———. *Report on the Gypsy Moth and the Brown-tail Moth*. U.S. Department of Agriculture, Bureau of Entomology, bulletin no. 58. Washington, D.C.: Government Printing Office, 1904.

———. "Pan American Cooperation in Plant Quarantine." *Proceedings of the Second Pan American Scientific Congress*, 888–904. Washington, D.C.: Government Printing Office, 1917.

———. *The San Jose Scale: Its Native Home and Natural Enemy*. U.S. Department of Agriculture. Washington, D.C.: Government Printing Office, 1902.

Marlatt, Charles Lester. *An Entomologist's Quest: The Diary of a Trip around the World, 1901–1902*. Baltimore: Monumental Printing Co., 1953.

———. "Pests and Parasites: Why We Need a National Law to Prevent the Importation of Insect-Infested and Diseased Plants." *National Geographic* 22, no. 4 (April 1911): 321–46.

———. "Protecting the United States from Plant Pests." *National Geographic* 40, no. 2 (August 1921): 205–18.

Mason, Silas C. "The Date in California." *California's Magazine* 1, no. 1 (July 1915): 470–76.

———. "Date Growing in Southern California." In *Official Report of the Thirty-Fourth Fruit-Growers Convention*, 170–78. Held April 28–May 1, 1908, Riverside. Sacramento, Calif.: W. W. Shannon, Superintendent of State Printing, 1908.

McClain, Cathi, "A New Look at Eliza Tibbets." *Citrograph*, October 1976, 449–53.

McCurdy, J.A.D. *Experiences in the Air*. May 23, 1908. In AEA bulletin no. 1, July 13, 1908, Alexander Graham Bell family papers, Library of Congress.

Meyer, Frank N. *Agricultural Explorations in the Fruit and Nut Orchards of China*. U.S. Department of Agriculture, Bureau of Plant Industry, bulletin no. 204. Washington, D.C.: Government Printing Office, 1911.

———. "Breeding for Horns." *Journal of Heredity* 6, no. 2 (February 1915): 96.

———. "China a Fruitful Field for Plant Exploration." U.S. Department of Agriculture, *Yearbook of Agriculture,* 205–24. Washington, D.C.: Government Printing Office, 1915.

———. "Collecting in Turkestan." *Journal of Heredity* 5, no. 4 (April 1914): 159–69.

———. "Economic Botanical Explorations in China." Lecture delivered at the Massachusetts Horticultural Society, March 25, 1916. Society archives, Wellesley, Mass.

———. "Seeking Plant Immigrants." *Journal of Heredity* 5, no. 3 (March 1914.) 111–21.

Mills, Cuthbert. *Letters to the New York Times during the Indian War of 1876.* La Mirada, Calif.: J. Willert, 1984.

Moon, David. "In the Russians' Steppes: The Introduction of Russian Wheat on the Great Plains of the United States of America." *Journal of Global History* 3, no. 2 (July 2008): 203–25.

Outhier, Craig, "George W. P. Hunt." *Phoenix Magazine,* December 2011. http://www.phoenixmag.com/History/george-w-p-hunt.html.

Page, Thomas Nelson. Unpublished biographical essay on Florence Lathrop Page. Page Collection, Special Collections, Alderman Library, University of Virginia, Charlottesville.

———. *Mediterranean Winter 1906: Journal and Letters.* Miami: Field Research Projects, 1971.

Pauly, Philip J. "The Beauty and Menace of the Japanese Cherry Trees: Conflicting Visions of American Ecological Independence." *Isis* 87, no. 1 (March 1996): 51–73.

———. *Biologists and the Promise of American Life: From Meriwether Lewis to Alfred Kinsey.* Princeton, N.J.: Princeton University Press, 2000.

———. *Fruits and Plains: The Horticultural Transformation of America.* Cambridge, Mass.: Harvard University Press, 2007.

"The People Who Stand for Plus." *Outing Magazine* 53, no. 1 (October 1908): 69–76.

Pickens, Donald K. *Eugenics and the Progressives.* Nashville: Vanderbilt University Press, 1968.

Poole, Robert M. *Explorers House: National Geographic and the World It Made.* New York: Penguin, 2004.

Popenoe, David. *Remembering My Father, Paul Popenoe: Portrait of the Man Who Saved Marriages.* New York: Institute for American Values, 1991.

Popenoe, F. O. "The San Francisco Disaster: A Personal Narrative." *Pacific Monthly,* June 1906, 727–36.

Popenoe, F. O., and T. U. Barber. *Tropical Fruits for California.* Altadena, Calif.: West India Gardens, 1911.

Popenoe, Paul, ed. "Constructive Eugenics." *Journal of Heredity* 5, no. 10 (October 1914): 458–62.

———. *Date Growing in the Old World and the New.* Altadena, Calif.: West India Gardens, 1913.

———. "Eugenical Sterilization: A Review." *Journal of Heredity* 14, no. 7 (October 1923): 308–10.

———. "Eugenics in Germany." *Journal of Heredity* 13, no. 8 (August 1922): 382–84.

———. "Genealogy and Eugenics." *Journal of Heredity* 8, no. 8 (August 1915): 372–83.

———. "The German Sterilization Law," *Journal of Heredity* 25, no. 7 (July 1934): 257–60.

———. "Heredity and the Mind." *Journal of Heredity* 7, no. 1 (October 1916): 456–62.

———. "Measuring Human Intelligence." *Journal of Heredity* 12, no. 5: 231–36.

———. "Nature or Nurture." *Journal of Heredity* 6, no. 5 (May 1915): 227–40.

———. "The Progress of Eugenic Sterilization." *Journal of Heredity* 25, no. 1 (January 1934): 19–25.

———. "An Unexpected Commission." *Desert Magazine*, March 1963, 46–49.

Popenoe, Wilson. "Avocado Explorations in Tropical America." *Proceedings of the Florida State Horticultural Society* 35 (1922): 31–36.

———. "David Fairchild, Plantsman." *Fairchild Tropical Garden Bulletin* 18, no. 3 (July 1963): 69–80.

———. "Hunting Avocados." *California Avocado Association Yearbook, Volume 17*, 180–82. California Avocado Society, 1932. http://www.avocadosource.com/CAS _Yearbooks/CAS_17_1932/CAS_1932_PG_180-182.pdf.

———. "Hunting New Fruits in Ecuador." *Natural History* 24, no. 4 (July–August 1924): 454–66.

———. "The Jaboticaba." *Journal of Heredity* 5, no. 7 (July 1914): 318–26.

———. *Manual of Tropical and Subtropical Fruits*. New York: Macmillan, 1920.

———. "Reflections, Reminiscences, and Observations." *California Avocado Association Yearbook, Volume 32*, 96–107. California Avocado Society, 1947. http://www. avocadosource.com/CAS_Yearbooks/CAS_32_1947/CAS_1947_PG_096-107. pdf.

———. "Round about Bogota." *National Geographic* 49, no. 2 (February 1926): 127–60.

Popenoe, Paul, and Roswell Hill Johnson. *Applied Eugenics*. New York: Macmillan, 1918.

Proceedings of the First National Conference on Race Betterment. Held January 8–12, 1914, Battle Creek, Mich. Race Betterment Foundation, 1914.

Punnett, Reginald Crundell. "More Knowledge Necessary." *Journal of Heredity* 5, no. 1 (January 1914): 33.

Quinn, Peter. "Race Cleansing in America." *American Heritage Magazine* 54, no. 1 (February/March 2003): 35–43.

Renwick, Edward A. *Recollections*. Evanston, Ill.: privately published, 2003.

Riddle, Oscar. "Charles Benedict Davenport, 1866–1944." *Biographical Memoirs of the National Academy of Sciences* 25, no. 4 (Autumn 1947). http://www.nasonline.org/ publications/biographical-memoirs/memoir-pdfs/davenport-charles.pdf.

Rixford, G. P. *Smyrna Fig Culture*. U.S. Department of Agriculture, Bureau of Plant Industry, bulletin no. 732. Washington, D.C.: Government Printing Office, 1918.

Robinson, T. Ralph. "Henry Perrine, Pioneer Horticulturalist of Florida." *Tequesta* 2 (1942): 16–24.

Rock, J. F. "Hunting the Chaulmoogra Tree." *National Geographic* 41, no. 3 (March 1922): 242–76.

———. "A Square Plug in a Round Hole." Unpublished autobiography. Joseph Francis Charles Rock papers. Arnold Arboretum, Harvard University, Jamaica Plain, Mass.

Roeding, George C. "Future of the Fig Industry." Official *Report of the Twenty-Ninth Fruit-Growers' Convention of the State of California*, 205–10. Held December 8–11, 1903, Fresno. Sacramento, Calif.: W. W. Shannon, Superintendent of State Printing, 1904.

———. "How the Smyrna Fig and Fig Insect Came to California." *Pacific Rural Press*, February 10, 1917, 164–65.

———. *The Smyrna Fig at Home and Abroad*. Fresno, Calif.: George C. Roeding, 1903.

Rogers, Andrew Denny, III. *Erwin Frink Smith: A Story of North American Plant Pathology*. Philadelphia: American Philosophical Society, 1952.

Roistacher, Chester N. "The Parent Washington Navel Orange: Its First Years." *Topics in Subtropics* (University of California Cooperative Extension newsletter) 7, no. 4 (October–December 2009): 4–7.

Rosengarten, Frederic, Jr. *Wilson Popenoe: Agricultural Explorer, Educator, and Friend of Latin America*. Lawai, Kauai, Hawaii: National Tropical Botanical Garden, 1991.

Royal Horticultural Society. "Hybrid Conference Report." *Journal of the Royal Horticultural Society* 24 (1900).

Ryerson, Knowles A. Transcript, unpublished, undated interview by Lois Stone. Ryerson records, Hunt Institute for Botanical Documentation, Pittsburgh.

———. "The History of Plant Exploration and Introduction in the United States Department of Agriculture." *Proceedings of the International Symposium on Plant Introduction*. Tegucigalpa, Honduras, Escuela Agricola Panamericana, November 30–December 2, 1966.

———. "Plant Introductions." *Agricultural History* 50, no. 1 (January 1976): 248–57.

———. *The World Is My Campus*. Davis, Calif.: Oral History Center, 1977.

———. "History and Significance of the Foreign Plant Introduction Work of the United States Department of Agriculture." *Agricultural History* 7, no. 3 (July 1933): 110–28.

Sasscer, E. R. "Charles Lester Marlatt." *Cosmos Club Bulletin*, September 1954, 2–5.

Schokman, Larry M. *Plants of the Kampong: A Guide to the Living Collection*. Coconut Grove, Fla.: National Tropical Botanical Garden, 2012.

Scidmore, Eliza R. "Capital's Cherry Blossoms Gift of Japanese Chemist." (*Washington*) *Sunday Star*, April 11, 1926.

———. "The Cherry Blossoms of Japan." *Century Magazine* 79, no. 5 (March 1910): 643–49.

Seifriz, William. "Walter T. Swingle: 1871–1952." *Science* 118, no. 3063 (September 11, 1953): 288–89.

Shamel, A. D. "The Parent Fuerte Avocado." *Journal of Heredity* 28, no. 5 (May 1937): 181–82.

———. "Washington Navel Orange." *Journal of Heredity* 6, no. 10 (October 1915): 435–45.

Simon, Hilda. *The Date Palm: Bread of the Desert*. New York: Dodd Mead, 1978.

Smith, Edwin F. "Frank N. Meyer," *Science* 48, no. 1240 (October 4, 1918): 335–36.

Spiro, Jonathan Peter. *Defending the Master Race: Conservation, Eugenics, and the Legacy of Madison Grant*. Burlington: University of Vermont Press, 2009.

Starr, Kevin. *Americans and the California Dream, 1850–1915*. New York: Oxford University Press, 1973.

Stern, Alexandra Minna. *Eugenic Nation: Faults and Frontiers of Better Breeding in Modern America*. Berkeley: University of California Press, 2005.

Stoner, Allan, and Kim Hummer. "19th and 20th Century Plant Hunters." *HortScience* 42, no. 2 (April 2007): 197–99.

Sutton, S. B. *In China's Border Provinces: The Turbulent Career of Joseph Rock, Botanist-Explorer*. New York: Hastings House, 1974

———. "Joseph Rock: Restless Spirit." In *Lamas, Princes, and Brigands: Joseph Rock's Photographs of the Tibetan Borderlands of China*, ed. Michael Aris, 22–26. New York: China House Gallery and China Institute in America, 1992.

Swanson, Arthur F. "Mark Alfred Carleton—The Trail's End." *Agronomy Journal* 50, no. 12 (1958): 722–23.

Swingle, Walter T. *The Date Palm and Its Utilization in the Southwestern States*. U.S. Department of Agriculture, Bureau of Plant Industry, bulletin no. 53. Washington, D.C.: Government Printing Office, 1904.

———. "The Dioecism of the Fig in Its Bearing upon Caprification." *Science* 10, no. 251 (October 20, 1899): 570–74.

———. "Our Agricultural Debt to Asia." In *The Asian Legacy and American Life*, ed. Arthur E. Christy, 4–15. New York: East and West Association, 1942.

———. "Some Citrus Fruits That Should Be Introduced into Florida." *Proceedings of the Sixth Annual Meeting of the Florida State Horticultural Society* 51 (1893): 111–22.

———. "Some Points in the History of Caprification in the Life History of the Fig." *Official Report of the Thirty-Fourth Fruit Growers Convention of the State of California*, 178–87. Held April 28–May 1, 1908, Riverside. Sacramento, Calif.: W. W. Shannon, Superintendent of State Printing, 1908.

Swingle, Walter T., and G. P. Rixford. "The First Establishment of Blastophaga in California." *Official Report of the Thirty-Eighth Fruit Growers Convention in the State of California*, 174–79. Held December 6–10, Stockton. Sacramento, Calif.: W. W. Shannon, Superintendent of State Printing, 1911.

Taylor, H. J. *To Plant the Prairies and the Plains: The Life and Work of Niels Ebbesen Hansen*. Mount Vernon, Iowa: Bios, 1941.

Tesche, W. C. "Rixford—Veteran Plantsman." *Journal of Heredity* 21, no. 3 (March 1930): 99–106.

Toward, Lilias M. *Mabel Bell: Alexander's Silent Partner*. Toronto: Methuen, 1984.

"Tribute to David Fairchild." *Fairchild Tropical Garden Bulletin*, November 10, 1954.

Troyer, A. F., and H. Stoehr. "Willet M. Hays, Great Benefactor to Plant Breeding and the Founder of Our Association." *Journal of Heredity* 94, no. 6 (2003): 435–41.

Tyson, Russell. *Report of Survey of Aldis and Company Records*. Chicago History Museum Research Center.

U.S. Department of Agriculture, Bureau of Plant Industry, Office of Seed and Plant

Introduction. *Inventory of Seeds and Plants Imported.* Nos. 1–76. Washington: Government Printing Office, 1898–1926.

———. *New Plant Immigrants.* Nos. 1–219. Washington, D.C.: Government Printing Office, 1908–24.

U.S. House of Representatives, 59th Cong., 2nd sess. *Hearings Before the Committee on Agriculture of Chiefs of Bureaus and Divisions and Other Officers of the Department of Agriculture on the Estimates of Appropriations for the Department of Agriculture for the Fiscal Year Ending June 30, 1908, Also of Seedsmen and Other Persons on Free Seed Distribution and Other Matters Relating to the Department of Agriculture.* Washington, D.C.: Government Printing Office, 1907.

———, 60th Cong., 1st sess. *Hearings Before the Committee on Agriculture of the Honorable Secretary of Agriculture and Chiefs of Bureaus and Divisions of the Department of Agriculture on the Estimates of Appropriations for the Fiscal Year Ending June 30, 1907, Also of Members of Congress and Other Persons Interested in Matters Pertaining to the Department of Agriculture and the Committee.* Washington, D.C.: Government Printing Office, 1908.

———, 61st Cong., 2nd sess. *Hearings Before the Committee on Agriculture on Miscellaneous Bills Including Inspection of Nursery Stock III.* Washington, D.C.: Government Printing Office, 1910.

———, 61st Cong., 3rd sess. *Quarantine against Importation of Diseased Nursery Stock.* House Reports 1, no. 1858. Washington, D.C.: Government Printing Office, 1911.

———, 62nd Cong., 2nd and 3rd sess. *Hearings Before the Committee on Agriculture on Miscellaneous Bills and Other Matters.* Washington, D.C.: Government Printing Office, 1912.

U.S. Senate, 76th Cong., 3rd sess. *Hearings Before the Committee on Agriculture and Forestry, April 25 and 27, 1940.* Washington, D.C.: Government Printing Office, 1940.

Van der Zee, John. *The Greatest Men's Party on Earth.* New York: Harcourt Brace Jovanovich, 1974.

Van Uildriks, Fredericke J. "De Aarde en Haa Volken." Article about Frank Meyer translated by Jeannette Bouter Bernaerts as "The Traveler-Botanist Frank N. Meyer and His Work," 1919. Cunningham collection, National Agricultural Library, U.S Department of Agriculture, Beltsville, Md.

Venning, Frank D. "Walter Tennyson Swingle, 1871–1952." *Carrell: Journal of the Friends of the University of Miami Library* 18 (1977): 1-33.

Walker, Egbert Hamilton. "Joseph F. Rock, 1884–1962." *Plant Science Bulletin* 9, no. 2 (April 1963): 7–8.

Walker, Hester Perrine. "Massacre at Indian Key, August 7, 1840, and the Death of Doctor Henry Perrine." *Florida Historical Society Quarterly* 5, no. 1 (July 1926): 18–42.

Wallace, Alfred Russel. "On the Bamboo and Durian of Borneo." Letter to Sir William Jackson Hooker, published 1856. http://people.wku.edu/charles.smith/wallace/S027.htm.

———. *The Malay Archipelago.* London: Macmillan, 1869. Singapore: Periplus, 2000.

Wallace, Henry A. News release. Speech awarding Meyer Medal to David Fairchild, July 24, 1939. Washington, D.C.: U.S. Department of Agriculture.

Walravens, Hartmut, ed. *Berichte, Briefe, und Dokumente des Botanikers, Sinologen und Nakhi-Forschers* [Reports, Letters, and Documents of the Botanists, Sinologist, and Naxi Researchers]. Stuttgart: F. Steiner, 2002.

———. *Joseph Franz Rock (1884–1962) Tagebuch der Reise von Chieng Mai nach Yunnan, 1921–1922; Briefwechsel mit C. S. Sargent, University of Washington, Johannes Schubert und Robert Koc* [Diary of a Journey from Chiang Mai to Yunnan, 1921–1922; Correspondence with C. S. Sargent, University of Washington, Johannes Schubert and Robert Koc]. Vienna: Verlag der Osterreichischen Akademie der Wissenschaften, 2007.

Watson, James. "Genes and Politics." In *Davenport's Dream: 21st Century Reflections on Heredity and Eugenics*, ed. Jan A. Witkowski and John R. Inglis, 1–34. Cold Spring Harbor, N.Y.: Cold Spring Harbor Laboratory Press, 2008.

Webber, H. J. "Explanation of Mendel's Law of Hybrids." *Journal of the American Breeders' Association* 1, no. 2 (January 1905): 138–43.

Weber, Gustavus A. *The Bureau of Entomology*. Washington, D.C.: Brookings Institution, 1930.

———. *The Plant Quarantine and Control Administration*. Washington, D.C.: Brookings Institution, 1930.

Wentzel, Volkmar K. "Gilbert Hovey Grosvenor, Father of Photojournalism." Cosmos Club of Washington, D.C., 1998. http://www.cosmosclub.org/web/journals/1998/wentzel.html.

Whitman, Walt. *Drum Taps*. New York, Grosset & Dunlap, n.d.

Wilcox, Earley Vernon, and Flora H. Wilson. *Tama Jim*. Boston: Stratford, 1930.

Wilder, Daniel W. *Annals of Kansas*. Topeka: G. W. Martin, 1875.

Wilder, Laura Ingalls. *On the Banks of Plum Creek*. New York: Harper and Brothers, 1937.

Wilkinson, Jerry. *Dr. Henry Edward Perrine*. Booklet written for Indian Key Festival, 1995. *Keys Historium*. Website hosted by Historical Preservation Society of the Upper Keys. http://www.keyshistory.org/Perrine-Dr-Page-1.htm.

———. "The Massacre Story." *Keys Historium*. Website hosted by Historical Preservation Society of the Upper Keys. http://www.keyshistory.org/IK-massacre-1.html.

Williams, Beryl, and Samuel Epstein. *Plant Explorer David Fairchild*. New York: Julian Messner, 1961.

Wilson, Ernest Henry. *Aristocrats of the Garden*. 1917. Carlisle, Mass.: Applewood Books, 2008.

———. *A Naturalist in Western China with Vasculum, Camera, and Gun*. Vol. 1. New York: Doubleday, Page, 1914.

———. "Charles Sprague Sargent." *Harvard Graduates Magazine*, June 1927, 614.

Wiser, Vivian D. *Protecting American Agriculture: Inspection and Quarantine of Imported Plants and Animals*. Agricultural Economic Report no. 266. Washington, D.C.: Economic Research Service, U.S. Dept. of Agriculture, 1974.

Woodger, Elizabeth R. "Wilson Popenoe, American Horticulturist, Educator, and Explorer." *Huntia* 5, no. 1 (1983): 17–22.

Woodger, Elizabeth R., and Arlyn Sharpe, comps.; Michael T. Stieber, ed. "An Inventory of the Wilson Popenoe Papers." *Huntia* 5, no. 1 (1983): 23–59.

Wulf, Andrea. *The Founding Gardeners: How the Revolutionary Generation Created an American Eden.* London: William Heinemann, 2011.

Zuckerman, Bertram. *The Dream Lives On: A History of the Fairchild Tropical Garden, 1938–1988.* Miami: Fairchild Tropical Botanic Garden, 1988.

———. *The Kampong: The Fairchilds' Tropical Paradise.* Kauai and Miami: National Tropical Botanical Garden and Fairchild Tropical Botanic Garden, 1993.

Writer Amanda Harris was for many years a reporter and editor at *Newsday*, the Long Island and New York City daily newspaper. She lives in Manhattan and Montcabrier, France, with her husband, Drew Fetherston. She is also the granddaughter of Alice Barton Harris and the daughter of Robin Franklin Harris, who was Marian Fairchild's godson.

proof